with a Bird,

A Reader on Avian Kinship

ONOMATOPEE 262

LEARNING THE NAME

for Bette

The wood thrush, it is!
Now I know
who sings that clear arpeggio,
three far notes
weaving into the evening
among leaves
and shadow;

or at dawn in the woods,
I've heard the sweet ascending
triple word
echoing over the silent river –
but never seen the bird.

Ursula K. Le Guin

INDEX

with a Bird,

A Reader on Avian Kinship

with a Bird, is the second iteration in the five-year unfolding of ***A Tree, with a Bird, by a Woman, on Land, Under a Star***, which aims to build space for rich encounters between folklore and critical research. The project title grows each year with the addition of each subsequent chapter, and it also builds and changes in meaning as the actors and points of focus shift. The title becomes a phrase that functions as a kind of spell, made of exit points or components that comprise their own ecology of stories and characters. Together, they form a composition, like a tarot card reading that speaks to the future and the past, that shares an insight, a warning, or a recipe for the now. Through a multitude of forms, the project commissioned by Onomatopee activates its multipurpose project space, morphing the outcomes and exchanges between exhibition, public program, publication, and workshop with a myriad of actors.

As thinking, feeling humans, we often place ourselves at the center of our own earthly narrative, as though the planet is and always has been ours. This is no surprise; as Homo sapiens, we've inhabited Earth for about 300,000 years. But our anthropocentrism is not as old as our species: informed by the binary oppositions of the Enlightenment—that separate nature and culture, animal and human, civilized and wild—the position of humankind as protagonist is actually a relatively recent one. Perhaps, then, the most potent antidote to our anthropocentrism can be found precisely in confronting time: fossils of the most recent common ancestor of modern birds date back a long 95 million years. Even further back we find the

Archaeopteryx, a more primitive bird species, who lived around 150 million years ago. Primates didn't appear until much later, and Homo sapiens later still. In other words, birds were here long before us, and our coexistence with them, particularly songbirds, may have shaped our lives in subtle but profound ways.

From here on the ground, we often hear birds more than we see them, and their songs are rich with both information and appeal. Birdsong moves us over physical and metaphysical planes, and presents a direct registration of life in a given moment and place. Take the blackbird, whose song is both learned and improvised—an artistic expression that communicates territory, identity, and emotion. Songbirds like the blackbird are an essential part of their environments in that they don't just sing *about* their surroundings, they literally sing them into shape. Some of their vocalizations attract mates, while other patterns—particularly the dawn and dusk choruses—synchronize with metabolic processes in nearby plants, optimizing their growth and preparation for nighttime. Exerting a tangible influence on the plant and animal life that thrives in its habitat, the blackbird's song is an integral element of what constitutes the ecosystem and its activity. The bird builds through sound.

Their songs also go on to reflect the ecosystems they co-create, leaving clues as to which other microorganisms and fungi, lichen, plants, and animals share the surroundings. Some birds are generalists, thriving in cities, while others require specific habitats, symbiotic or interspecies relationships, and diets. But wherever birds sing, they make audible the richness of the land—not just its geography, but the dynamic web of life it holds: the microbes in the soil, the plants that feed the insects, the insects that feed the chicks, and the weather that shapes it all. Through their song, birds become translators of biodiversity, offering a musical map of our shared environment.

We can also learn a lot from sound that isn't there; one intriguing example of avian influence on humans is how we interpret silence. Songbirds sing only in environments free of major threats. The moment danger approaches, they fall quiet. This silence can signal to us that something is wrong, and over time, we may have evolved to tune into this cue. Walking through a silent forest or woodland, for example, can feel unnerving, while in popular culture, characters with hidden motives are often depicted whistling—an eerie mimicry of birdsong that tries to conceal their ill intentions. Perhaps our ability to perceive birdsong as pleasant is more than aesthetic; it could be a survival trait, refined through millennia of shared habitats. Beyond its relation to survival, birdsong also nurtures the human mind. Scientific studies show that hearing birds improves mental health, alleviating symptoms of depression, anxiety, and paranoia. At metro stations in Tokyo, recordings of cuckoo calls and other songbirds guide visually impaired passengers to the platforms and promote a sense of calm for the throngs of commuters.

Whether in the city or the wild, birds voice the vitality and fragility of our interconnected systems, but many of those voices have already gone quiet. Climate breakdown and habitat destruction, symptoms of the extractive practices of the Capitalocene, have driven countless species to decline or extinction, unraveling the ecological relationships that once sustained them. Their silence is not just absence—it is a warning, echoing through the diminishing chorus of the living world. By recognizing birdsong and the information and needs it expresses, we gain insight into the life around us. There's still so much we don't understand about our winged companions, but perhaps listening closely is the best place to start. Learning to listen to them is, in essence, learning to understand nature as our home and ourselves.

For centuries, even the basic movements of birds were misunderstood, reminding us how often human knowledge begins in mystery and is shaped by wonder. Until two hundred years ago, scientists in Europe widely believed that birds spent the autumn and winter months on the moon, living underwater, dwelling in swamps, or otherwise hiding in holes and crevices. These imaginative theories were generally accepted until a remarkable discovery shattered any need for them. In 1822 in Germany, a stork that had been shot down mid-flight appeared to already have a long arrow fast in their neck. The first arrow was traced back to Central Africa, revealing that the bird had traveled over 6,000 kilometers while surviving the injury. Instead of spending the winters hiding from the cold, it became clear that these birds were living on the move, their home changing with the seasons. This incident revolutionized our understanding of avian behavior and migration.

Many migratory birds—including the black-tailed godwit, northern lapwing, common nightingale, willow warbler, blackcap, chaffinch, and song thrush—enliven European gardens, fields, forests, wetlands, and coastlines with their songs during spring and summer. These same birds then move to warmer climates during winter, a season in which food becomes scarce, yet when cold and wet conditions demand extra energy to stay warm. For humans, winter has historically been a time of introspection. With shorter days and less sunlight, people often sleep more, retreat into their homes, and conserve energy for essential tasks. During these darker months, much of outdoor life dwindles, and people tend to spend time together indoors, perhaps around a fire, sharing all kinds of stories. Winter is a season of reflection, storytelling, and imagination. In most of Europe, it is also a season without many migratory birds, but culturally, they are never gone; birds have long captured human fascination with their ability to fly, navigate vast distances, and produce melodious songs. They play a significant role in our cultural and spiritual imaginations, inspiring myths, folk tales, and scientific curiosity. The desire to emulate birds—whether through airplanes, navigation systems, or musical instruments—has profoundly shaped technological advancements and philosophical ideas. Birds continually challenge and expand our understanding of the world, defying the boundaries and categories that we humans not only construct but are bound to.

The exhibition **with a Bird,** taking place from mid-January to late April 2025, is the second expression of *A Tree, with a Bird, by a Woman, on Land, Under a Star.* It showcases eight projects, objects, and investigations through which the artists, Ignace Cami, Monika Czyżyk, Bryony Dunne, Daniel Godínez Nivón, Manjot Kaur, Ai Ozaki, Sergio Rojas Chaves, and Sara Sejin Chang (Sara van der Heide), explore our relationships with birds. These works delve into the ways we seek to understand, emulate, and connect with birds, while examining how avian life transcends categories such as human and non-human; science and folklore; life and death; reality and dreams; and the realms of land, water, and sky. As part of the **with a Bird,** project, this publication can be seen as a continuation of the exhibition. It makes a different kind of space for aspects of the work and research of each artist to sit alongside one another, creating new configurations, associations, and lines of dialogue. It invites the reader, at their own pace, to spend time with intimate experiences, personal stories, scientific backgrounds, and historical contexts of human–bird relations and depictions across time and cultures. Throughout this reader, the aim is to nurture and further these dialogues and to share inspiration on exercising avian kinship by taking the time to think about birds differently, through imagination, art, storytelling, poetry, and images. Moreover, the contributions inspire us to consider overarching systems of oppression by zooming in and looking closely at our relations and surroundings. Through this focused engagement, we can move towards exorcizing the destructive forces of anthropocentrism, capitalism, individualism, heteronormativity, and coloniality.

The artists and their practices presented in the exhibition are the central framework for this edited volume. Structured in five sections, the book invites readers to explore the intersections of art, science, fiction, and storytelling through a diverse array of perspectives.

HEIGHTENING THE SENSES

The first section serves as a primer, laying the thematic foundations for the book's interweaving of perspectives, stories, and observations on avian–human–planetary relations. From making connections between looking and artmaking to confronting questions of mass extinction, the texts in ***Heightening the Senses*** set the stage, laying bare systems of destruction as well as daring to imagine and propose hopeful alternatives.

In *The Second Body*, Daisy Hildyard begins with the story of caring for a wounded pigeon—an intimate, visceral encounter that becomes a metaphor for her central idea: that each person exists as two selves—a physical, local body, and a second, planetary body, whose actions ripple invisibly across ecosystems. Through her luminous prose, Hildyard argues that only by emotionally connecting with this abstract second self can we begin to grasp our responsibility within a globalized, interconnected world.

In *The White Bird*, John Berger reflects on a simple wooden dove crafted by peasants as a symbol of beauty, mystery, and hope. He argues that art, like this bird, transforms fleeting moments of recognition in nature into enduring forms of meaning, offering resistance and affirmation in a world marked by suffering.

In *The Whiteness of Birds*, Nicholas Mirzoeff examines how ornithology, birdwatching, and visual culture have historically supported settler colonialism and racial capitalism. Focusing on figures like John James Audubon, he reveals how birds became tools of imperial classification and exclusion. Mirzoeff contrasts this with the concept of murmuration—collective, fluid movement—as a metaphor for decolonial resistance and alternative, relational ways of living and perceiving.

In *The Silent Academy / Chapter 2: The Lyrebird* by Yuri Tuma, we find ourselves in a dystopian future, where spoken language is suppressed and bird communication is forbidden. We follow a young rebel trained by the Silent Academy as they navigate a ruined forest in search of their companion, Joy, guided by whistled language and ancestral memory. Amid black-noise warfare and sonic ruins, a lyrebird's mimicry transforms destruction into defiance, while an embedded meditative exercise invites the reader to imagine across extinction and emergence, past and future, through the lens of avian kinship.

DEATH, DREAMS, AND OTHER REALMS

The second section, titled ***Death, Dreams, and Other Realms***, delves into the liminal world that birds connote. We follow the images and imaginations that are rooted in the everyday and branch out into other dimensions in all possible directions.

Belly Full of Worms is a site-specific series of ephemeral clay paintings on glass, created by Monika Czyżyk. Using earth pigments collected by the artist, the works depict birds and symbols that explore impermanence, transformation, and the porous boundaries between life and death, presence and absence. Through documentation of the work by Nick Bookelaar and an accompanying text, the works—guided by mythology, memory, and intuitive process, and infused with references from illuminated manuscripts and personal encounters—invite viewers to reflect on cycles of decay and renewal, and the lingering traces left by fleeting moments and more-than-human kin.

Drawing on recent neuroscientific research, in *Do Birds Dream?*, Maria Popova explores how birds, especially songbirds, rehearse melodies and flight patterns in REM sleep, revealing that avian dreams may mirror human cognitive and emotional processing. Suggesting that the capacity to dream may have first evolved in bird brains, she poetically argues that dreaming is a practice of possibility—one that links birds and humans across time through the rehearsal of memory, emotion, and imagined futures.

The essay *Dreamwind* by Daniel Godínez Nivón blends cosmology, memory, and ancestral knowledge. This poetic meditation traces the Earth's primordial dreaming into the dream-songs of birds, suggesting that their songs carry geological memory and offer portals into both past extinctions and future possibilities. Through personal narrative and collective oneiric practices, the text invites us to listen deeply, recognizing birdsong as both an ancient inheritance and a vital act of resistance, care, and world-making.

Sara Sejin Chang (Sara van der Heide)'s *The Spirit* is presented through documentation accompanied by a short text on the work. The oil painting, made with translucent layers that assemble the physical and the spiritual, draws from a newspaper image to portray a haunting, metaphysical embrace between figures. The work invites reflection on ancestry, spectral presence, and transformation. *The Spirit* anchors the human–bird relationship in liminal space, offering a vision of interconnected cosmologies.

IDENTITY AND IMAGINATION

The third section, *Identity and Imagination*, examines the role of birds in shaping human perceptions of self, culture, and creativity. In *How to Cockatoo*, artist Sergio Rojas Chaves examines how non-native species like the cockatoo enter Costa Rica's national imaginary through mass-produced souvenirs shaped by global tourism and neo-colonial consumption. By transforming everyday objects into cockatoo-inspired sculptures, he critiques the ways that national identity can be distorted by the exoticizing gaze, revealing how images divorced from ecological or cultural roots can become symbols of manufactured authenticity.

In the essay *The Greener Green*, editor and author Marjolein van der Loo traces their growing obsession with the Nordic nightingale, whose mysterious, shape shifting song evokes imagined jungles, ecological entanglements, colonial histories, and mythic storytelling. As birdsong blurs the boundary between dream and reality, the nightingale becomes a "strange stranger"—a companion and muse—through whom the writer explores themes of nature, memory, migration, and the power of listening as a way to reimagine our place in a fragile, interconnected world.

In *The Sociality of Birds*, anthropologist Anna Lowenhaupt Tsing explores how humans and birds co-create complex social worlds through interactions shaped by infrastructure, belief systems, and economic aspirations, focusing on how birdwatching practices intersect with local histories and ontologies. The text is situated in Waigeo, Indonesia, where Tsing examines "edge effects"—zones of unexpected contact between world-making projects. Here, she reveals the potential for multispecies collaboration and mutual curiosity, challenging colonial binaries and inviting a rethinking of interspecies relationality.

Artist Ai Ozaki's visual essay *Drawings* draws from everyday encounters and personal history, using birds—particularly pigeons and parakeets—as recurring figures through which she explores themes of home, bodily perception, and the fluidity of identity. Her intimate works reflect a nuanced engagement with the complexities of gender, sexuality, and cultural conditioning, rendering birds as both symbolic and deeply personal interlocutors.

Lastly, an excerpt from the scientific article by ornithologists Michelle J. Moyer, Evangeline M. Rose, Bernard Lohr,

Karan J. Odom, and Kevin E. Omland, here retitled as *Challenging Ornithological Bias: Female Song in Orchard Orioles*, presents the first comprehensive analysis of female song in orchard orioles, revealing that females sing more frequently than previously believed, and that their songs are structurally distinct from those of males. These findings challenge longstanding assumptions in ornithology by highlighting the importance of documenting female vocalizations to fully understand the evolution and function of birdsong across sexes.

FOLKLORES AND FUTURES

The fourth section, *Folklores and Futures*, draws on a variety of rich storytelling traditions to reflect on the longstanding existence of biodiverse, interspecies experience. *Tallio,* by artist Ignace Cami, tells the story of an art student whose creative awakening is catalyzed by her extraordinary bond with a talkative Eurasian jay, whose presence transforms her life, art, and identity. Through enchantment, jealousy, and loss, the story explores the transformative power of interspecies kinship and the enduring, ethereal trace of inspiration that remains long after the muse is gone.

Hybrid Beings by Manjot Kaur is an artistic project presented here with images of Kaur's original painted works, depicting bird–woman hybrids, and expressed in a poetic statement that reimagines romantic, ecological, and spiritual relationships between human and non-human life. Rooted in feminist, post-human, and post-queer thought, it challenges hierarchical, anthropocentric, and heteronormative systems, proposing a speculative world of multispecies kinship, ecological repair, and love as a transformative, interspecies force.

In *The Art of Deception*, artist Bryony Dunne recounts her 2023 residency in Askeaton, Ireland, where she installed sculptural bird decoys in the polluted River Deel to reflect on myth, ecology, and the histories of deception embedded in hunting, colonialism, and conservation. In response, in the text *Entrapping the Eye*, Suzanne Walsh, an artist and writer with roots along the River Slaney, weaves her own reflections on local duck hunters, camouflage, folklore, and the uncanny power of decoys, highlighting their shared ability to lure, mislead, and reveal. Together, their texts form a layered dialogue on deception as both artistic method and ecological mirror, where decoys become mediators between human perception, environmental urgency, and mythic imagination.

The essay *Ducks into Houses* by Marianne Elisabeth Lien explores the eider duck nesting practices on the Vega Archipelago in Northern Norway as a unique form of human–animal relationship that challenges conventional ideas of domestication. Instead of dominance or confinement, these relations are based on trust, care, and mutual benefit; eiders choose to nest in human-made shelters, and in return, people gather their down. This seasonal, multispecies cohabitation illustrates an alternative mode of "co-domestication" that is relational, situated, and shaped by precarious, lived entanglements.

CODA: ECHOES AND OMENS

Lastly, the book finishes with the essay *Fallen Angels: Birds of Paradise in Early Modern Europe* by Natalie Lawrence, which functions as a coda, a resonant final note. This richly researched piece traces how early European encounters with birds of paradise—arriving legless and lifeless from distant colonies—led to wildly imaginative theories: that these birds never landed, nourished only by air and sunlight, drifting forever between heaven and Earth. These interpretations, shaped by awe and imperial distance, remind us how the boundaries between myth, science, and power have long been porous. In dialogue with one of the book's opening stories—of a wounded stork whose body revealed a 6,000-kilometer migratory journey—Lawrence's text returns us to the entanglement of human wonder and avian movement. As the final entry in a project that resists linear time and treats storytelling as a spell of interwoven pasts and speculative futures, this essay becomes both echo and provocation. It asks what we inherit from outdated scientific imaginaries, and what new relationships between humans and birds might still take flight.

Alongside these contributions, a selection of images and artworks from the archives of Gallica (France), the Smithsonian Institution and The Metropolitan Museum of Art (United States), the Rijksmuseum and the Nationaal Archief (The Netherlands), and Europeana (Europe-wide) provide some historical and visual context. Please note that all of these collections were founded from a colonial, white, and Christian perspective—a viewpoint that sits at the root of how Western society (mis)treats much of life on Earth. However, the book aims to add a critical perspective, nurture a collaborative mindset, and contribute to a growing awareness of this history.

As this book opens with a story of rediscovery—of migration, mystery, and human humility—it also invites a return to awareness in the everyday. There are small, effective, accessible ways to support bird life that begin with attention and care. Listening closely, not just to birdsong but also to its absence, can reveal quiet losses unfolding around us. Staying informed through resources like *The State of the World's Birds*, an online resource reporting on bird life, helps us connect local experience with global patterns. As well as learning, we can act, in simple ways, from our own homes. Grow native plants. Refrain from using insecticides and avoid releasing balloons. Keep cats indoors or outfit them with bells and colorful collars to give birds a crucial advantage. Reduce light pollution by turning off unnecessary lights at night, or draw the curtains to prevent disorientation in migrating species. Even providing safe, pesticide-free food or water can help make shared spaces more hospitable. These gestures, while modest, cultivate a deeper practice of coexistence—one that extends the spirit of this book into everyday life.

In this spirit of attunement and reciprocity, the following voices offer stories, studies, and visions that deepen our understanding of birds—and of ourselves.

I very proudly introduce you to: John Berger, Ignace Cami, Sara Sejin Chang (Sara van der Heide), Monika Czyżyk, Bryony Dunne and Suzanne Walsh, Daniel Godínez Nivón, Daisy Hildyard, Manjot Kaur, Natalie Lawrence, Marianne Elisabeth Lien, Michelle J. Moyer, Evangeline M. Rose, Bernard Lohr, Karan J. Odom, Kevin E. Omland, Nicholas Mirzoeff, Ai Ozaki, Maria Popova, Sergio Rojas Chaves, Anna Lowenhaupt Tsing, Yuri Tuma.

Marjolein van der Loo

Primer:

Heightening the Senses

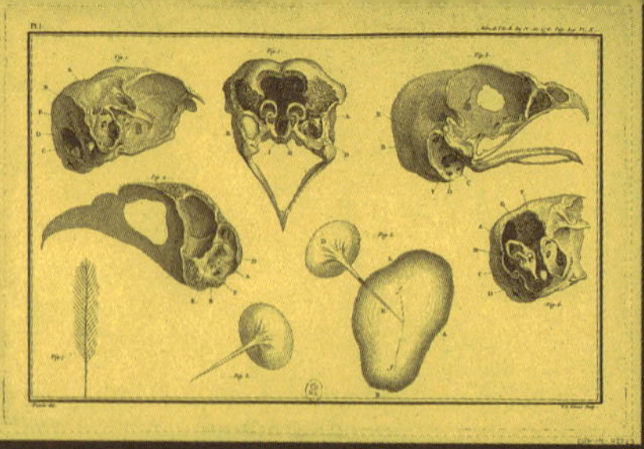

1)

(2)

6)

(7)

11)

(12)

(13)

(17)

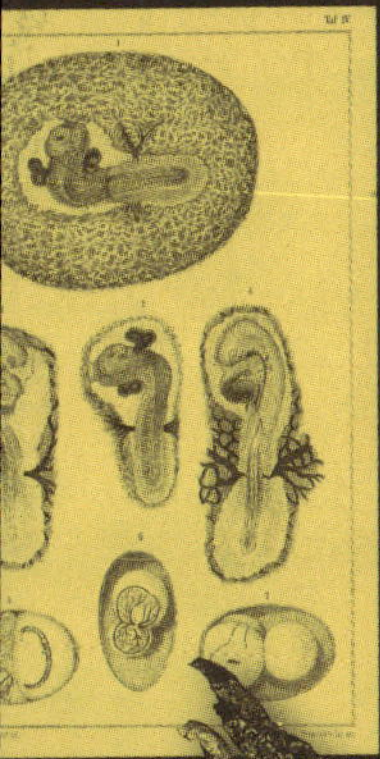

(4)

(5)

(8)

(9)

(10)

(15)

(20)

(21)

(1

(6)

(11)

(16)

(17)

(18)

(20) (21)

(1)

(6)

(7)

(12)

(13)

(11)

(16)

(17)

(18)

(8)

(15)

(20)

N°: 1341 18.-

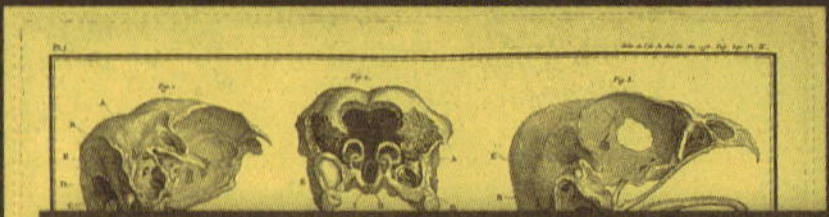

(1

(6

(11)

(17)

(16)

(18)

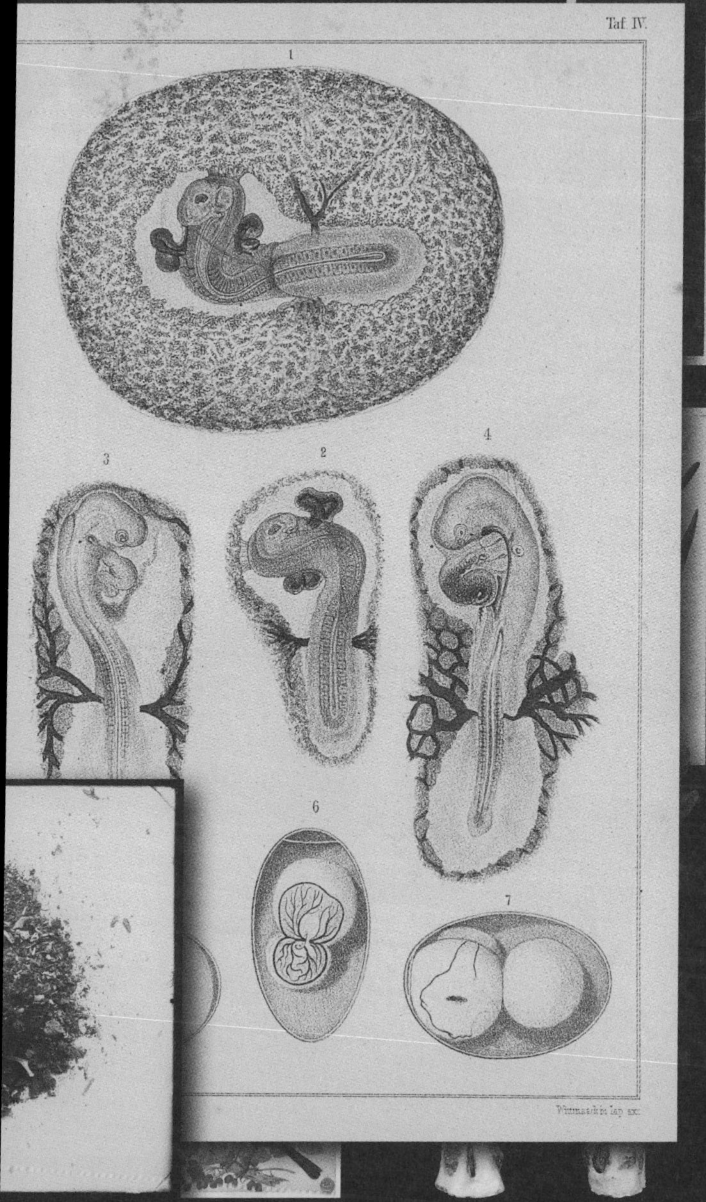

(20) (21)

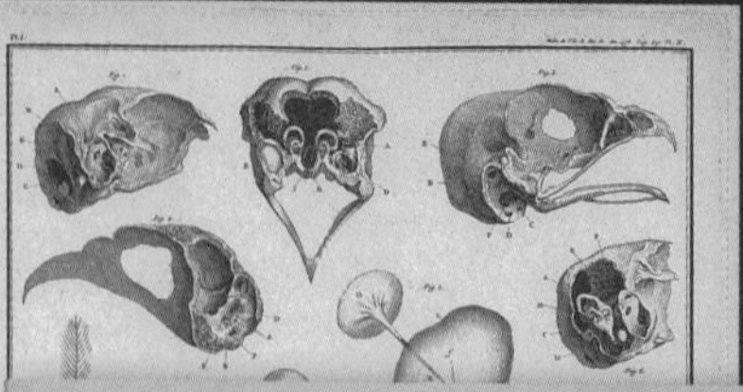

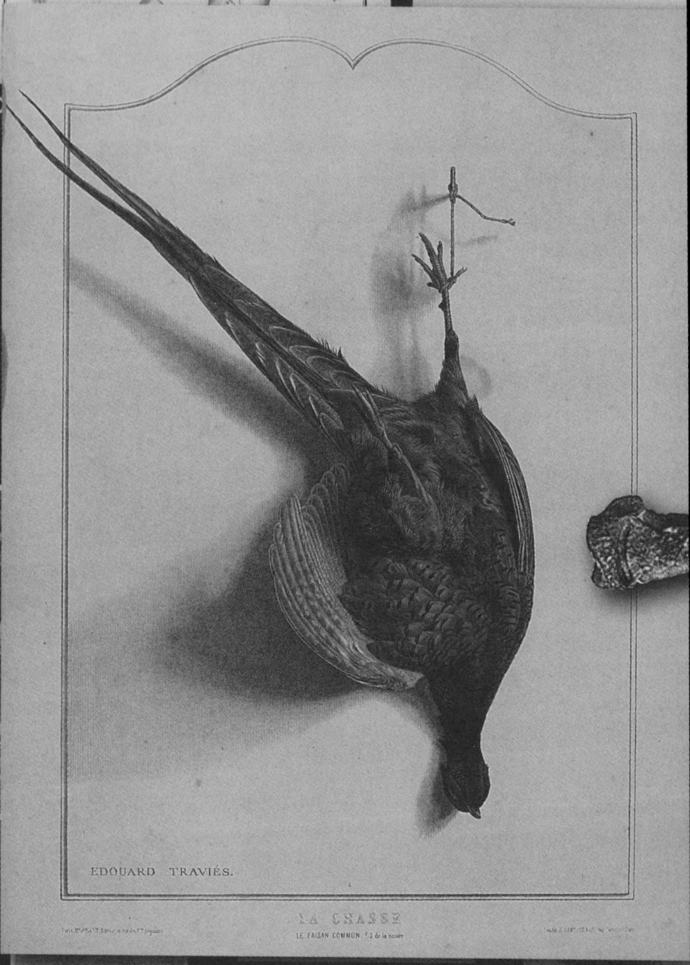

(17)

16)

(18)

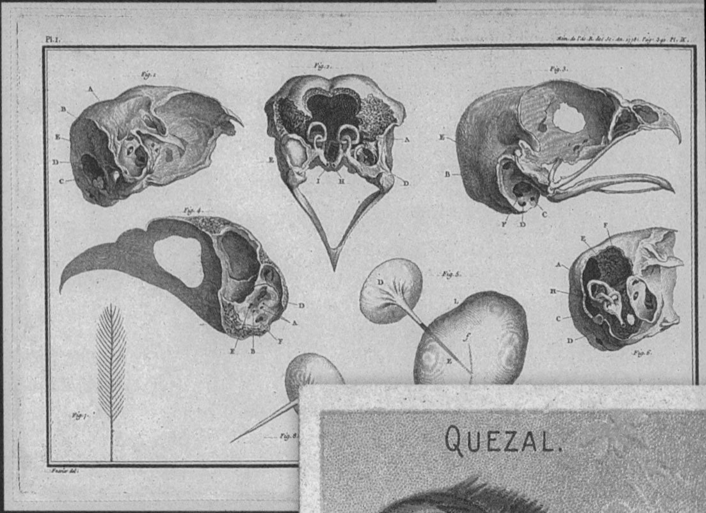

(6)

(11)

(12)

(17)

(16)

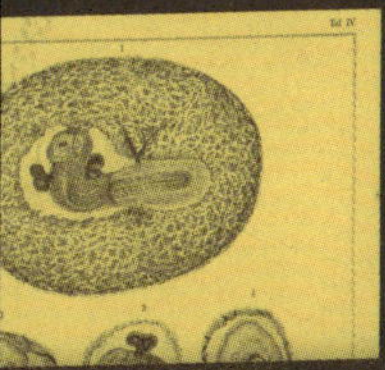

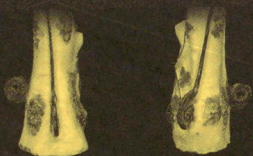

(20)

(21)

THE SECOND BODY
BY DAISY HILDYARD

I was alone in my kitchen when I noticed a small brown pigeon on the floor. The pigeon made a squeaking sound when I approached it, and I realized it couldn't fly. I crouched down until I was close enough to see the threads of pale fluff sticking out of its neck feathers: it must have been young. I tried to catch it but it was slipperier than I'd have expected – it went out of my hands like a fish. We stood looking at each other for a short while, and then I tried again. For a few seconds I had it in my grasp, and it did occur to me then that I could have wrung its neck and eaten it but I didn't – I put it in the shed, and cared for it for a while. It was quite greedy and fell off the edge of a bucket once while trying to get at seed. Its legs leaned at an angle over its claws: *L*. The talons curved up out of the ends of each toe and didn't seem to actually touch the ground when it walked. It was definitely using its eyes – I could only look at one of its eyes at a time. A very round pale brown eye which blinked. I could see its mind in its body. It turned its head every few seconds to get a proper view of me.

I fed my pigeon every morning for several weeks. It still wouldn't fly. I was worried it would be attacked by a rat – we've had rats in the yard – and one day I lost patience. I thought: it has to live for itself. I turned my pigeon outside and closed the shed door. The pigeon spent hours scuffling around on the concrete, and I nearly went out to collect it and let it back in, but then suddenly it was up on the roof. I didn't see it get there, and it rested for a while. Perhaps it will fall, I thought, but I was watching when two other pigeons, both grey, both larger than my pigeon, came for it. They were all on the grey roof together and then they all, at the same time, opened their wings and floated into the air, incredibly slowly. They were close together, and their wings made a kind of dome shape from where I was standing. It didn't look like they were flying so much as it looked like they were being pulled into space. I could see the three of them lifted further, and eventually they were so far away that I couldn't see them. It was December and the sky was almost white. Afterwards I went to the sink and cleaned and dried my hands. It was strange that the pigeon had previously been constrained by them.

It was around the same time that I started noticing the other animals. In the newspaper I saw images of eleven hippopotami which floated, dead, down a river in Binga, Zimbabwe. Dozens of barn owls fallen onto the Interstate-84 in Idaho. Tonnes of fish silvering the beaches in Montevideo, Uruguay. Hundreds of reindeer strewn across a plateau in southern Norway, after a freak storm.

Closer to home, I saw images of the corpses of sperm whales which had washed up on the coast at Skegness in England, some way to the south of where I live. I saw photographs of the bodies of pilot whales which had washed up, a few weeks later, on the coast in Fife in

Scotland, some way to the north of where I live. I saw images of winter-resident waxwings arriving at a wetland reserve not far from my house. Usually only a few hundred waxwings settle there for the season, but this last year thousands and thousands and thousands of them came – when they appeared over the horizon, the sky grew dark. In the summer, a study published in the journal Nature reviewed 370,000 ecological records from 1960 to 2012, and found that the seasons themselves were slowly drifting out of place.

While my small brown pigeon was flying away from me, it felt like these strange and prodigious animals were coming closer. I hadn't really noticed them before and then suddenly they seemed to be all over the place, landing on the beaches or bombing out of the sky. But still, none of these animals felt like they had anything to do with my pigeon because they just weren't as real. Their strange behaviour was more like a representation of something and I found it difficult to put my finger on what exactly it was. I looked online at shots of chromosomes, hydrocarbon data, satellite images. I watched impressions of pure white ice shelves breaking off and floating away. But all those things, like the owls and the Uruguayan fish, seemed to embody a truth which felt conceptual or abstract to me. The pigeon, on the other hand, even if it was flightless and unable to live on its own, even if the squeaking sound it made was sometimes annoying, and could not realistically be called a coo, even so, it was definitely there and I had become involved with it. My pigeon and I found time for one another every day. In my job, I speak to people who work with animals, and they, too, often give me an impression that there are different ways to exist in a body: that there are truths about any body – your body – which are not quite the same as the reality of that body's everyday life. The things I know to be true, in an abstract sense: satellite images, shots of chromosomes, hydrocarbon spreadsheets – they don't always feel real. Meanwhile, the real, fleshy, living bodies, going about their business, cleaning their kitchens or trying to get at their seed, falling off their buckets – they don't feel like they have much to do with the complicated truth about what is happening to life on earth. I find it hard to make myself much interested in this truth – it feels far off. I don't want to hear about climate change or the biosphere, I want to hear about real people and real creatures. But there is a sense that the sky is getting dark and the horizon is moving nearer – that I *should* be paying attention, because one day the distant ice shelf will come ripping through the tissue of my body – through every body – even if it appears, for now, that the bodies all around me are intact.

What does all this have to do with you? Everything. What do an American barn owl, a Zimbabwean hippopotamus and a Norwegian reindeer have in common? What they have in common is that they all have a relationship with your body – they are all, in some sense, your responsibility. There is a way of speaking which implicates your body in everything on earth. Dead whales have something to do with you, the disorientation of the waxwing is indirectly your problem, the freak storm and the changing seasons are consequences of actions performed by your body. Meanwhile, in the human world, there are car bombs going off in Baghdad every day. Does this have anything at all to do with you? Moreover, a teenager in Kolkata is missing a thumb and you are wearing a pair of inexpensive gloves. Is there any connection there?

The idea that a human body can be responsible for something which bears no tangible relation to it or to its immediate surroundings is not a new idea. In the scriptures, God sends down plagues and floods when men are misbehaving. In Macbeth, strange things that happen in nature, which seem at first to be spooky and supernatural, end up being logical – too logical, really – technical and fussy. When the horses eat each other in the stables, there is a feeling that something is about to go wrong; it does. The witches predict that *none of woman born* will harm Macbeth, and that he will be undefeated until Great Birnam Wood, to high Dunsinane Hill / Shall come against him. Later, he is killed by Macduff, who was from his mother's womb / Untimely ripped – and therefore not, in the phrasing of the time, *of woman born.* The forces that come to depose Macbeth come hidden behind branches which they lop off trees in Great Birnam Wood to take with them on the advance to Dunsinane Hill. Everything that happens, political and natural, is an effect of human acts, but the reasons are so obscenely down-to-earth that it takes a leap of imagination to perceive it.

This idea of a body which can reach over to the other side of the world is not one we tend to speak of in everyday language right now. In normal life, a human body is rarely understood to exist outside its own skin – it is supposed to be inviolable. The language of the human animal is that of a whole and single individual. You are encouraged to be yourself and to express yourself – to be whole, to be one. Move away from this personality, self-expression, and you risk going out of your mind, being beside yourself, failing to be true to yourself, hearing other voices or splitting your personality: it doesn't sound good. This careful language is anxious, I think – threatening in a desperate way. You need to take care of yourself, it says. You need boundaries, you have to be either here or there. Don't be all over the place.

Climate change creates a new language, in which you have to be all over the place; you are always all over the place. It makes every animal body implicated in the whole world. Even the patient who is anaesthetized on an operating table, barely breathing, is illuminated by surgeons' lamps which are powered with electricity trailed from a plant which is pumping out of its chimneys a white smoke that spreads itself out against the sky. This is every living thing on earth.

The White Bird

John Berger

From time to time I have been invited by institutions – mostly American – to speak aesthetics. On one occasion I considered accepting and I thought of taking with me a bird made of white wood. But I didn't go. The problem is that you can't talk about aesthetics without talking about the principle of hope and the existence of evil. During the long winters the peasants in certain parts of the Haute Savoie used to make wooden birds to hang in their kitchens and perhaps also in their chapels. Friends who are travellers have told me that they have seen similar birds, made according to the same principle, in certain regions of Czechoslovakia, Russia and the Baltic countries. The tradition may be more widespread.

The principle of the construction of these birds is simple enough, although to make a fine bird demands considerable skill. You take two bars of pine wood, about six inches in length, a little less than one inch in height and the same in width. You soak them in water so that the wood has the maximum pliability, then you carve them. One piece will be the head and body with a fan tail, the second piece will represent the wings. The art principally concerns the making of the wing and tail feathers. The whole block of each wing is carved according to the silhouette of a single feather. Then the block is sliced into thirteen thin layers and these are gently opened out, one by one, to make a fan shape. Likewise for the second wing and for the tail feathers. The two pieces of wood are joined together to form a cross and the bird is complete. No glue is used and there is only one nail where the two pieces of wood cross. Very light, weighing only two or three ounces, the birds are usually hung on a thread from an overhanging mantelpiece or beam so that they move with the air currents.

It would be absurd to compare one of these birds to a Van Gogh self-portrait or a Rembrandt crucifixion. They are simple, home-made objects, worked according to a traditional pattern. Yet, by their very simplicity, they allow one to categorize the qualities which make them pleasing and mysterious to everyone who sees them.

First there is a figurative representation – one is looking at a bird, more precisely a dove, apparently hanging in mid-air. Thus, there is a reference to the surrounding world of nature. Secondly, the choice of subject (a flying bird) and the context in which it is placed (indoors where live birds are unlikely) render the object symbolic. This primary symbolism then joins a more general, cultural one. Birds, and doves in particular, have been credited with symbolic meanings in a very wide variety of cultures.

Thirdly, there is a respect for the material used. The wood has been fashioned according to its own qualities of lightness, pliability and texture. Looking at it, one is surprised by how well wood becomes bird. Fourthly, there is a formal unity and economy. Despite the object's apparent complexity, the grammar of its making is simple, even austere. Its richness is the result of repetitions which are also variations. Fifthly, this man-made object provokes a kind of astonishment: how on earth was it made? I have given rough indications above, but anyone unfamiliar with the technique wants to take the dove in his hands and examine it closely to discover the secret which lies behind its making.

These five qualities, when undifferentiated and perceived as a whole, provoke at least a momentary sense of being before a mystery. One is looking at a piece of wood that has become a bird. One is looking at a bird that is somehow more than a bird. One is looking at something that has been worked with a mysterious skill and a kind of love.

Thus far I have tried to isolate the qualities of the white bird which provoke an aesthetic emotion. (The word 'emotion', although designating a motion of the heart and of the imagination, is somewhat confusing for we are considering an emotion that has little to do with the others we experience, notably because the self here is in a far greater degree of abeyance.) Yet my definitions beg the essential question. They reduce aesthetics to art. They say nothing about the relation between art and nature, art and the world.

Before a mountain, a desert just after the sun has gone down, or a fruit tree, one can also experience aesthetic emotion. Consequently we are forced to begin again – not this time with a man-made object but with the nature into which we are born.

Urban living has always tended to produce a sentimental view of nature. Nature is thought of as a garden, or a view framed by a window, or as an arena of freedom. Peasants, sailors, nomads have known better. Nature is energy and struggle. It is what exists without any promise. If it can be thought of by man as an arena, a setting, it has to be thought of as one which lends itself as much to evil as to good. Its energy is fearsomely indifferent. The first necessity of life is shelter. Shelter against nature. The first prayer is for protection. The first sign of life is pain. If the Creation was purposeful, its purpose is a hidden one which can only be discovered intangibly within signs, never by the evidence of what happens.

It is within this bleak natural context that beauty is encountered, and the encounter is by its nature sudden and unpredictable. The gale blows itself out, the sea changes from the colour of grey shit to aquamarine. Under the fallen boulder of an avalanche a flower grows. Over the shanty town the moon rises. I offer dramatic examples so as to insist upon the bleakness of the context. Reflect upon more everyday examples. However it is encountered, beauty is always an exception, always *in despite of*. This is why it moves us.

It can be argued that the origin of the way we are moved by natural beauty was functional. Flowers are a promise of fertility, a sunset is a reminder of fire and warmth, moonlight makes the night less dark, the bright colours of a bird's plumage are (atavistically even for us) a sexual stimulus. Yet such an argument is too reductionist, I believe. Snow is useless. A butterfly offers us very little.

Of course the range of what a given community

finds beautiful in nature will depend upon its means of survival, its economy, its geography. What Eskimos find beautiful is unlikely to be the same as what the Ashanti found beautiful. Within modern class societies there are complex ideological determinations: we know, for instance, that the British ruling class in the eighteenth century disliked the sight of the sea. Equally, the social use to which an aesthetic emotion may be put changes according to the historical moment: the silhouette of a mountain can represent the home of the dead or a challenge to the initiative of the living. Anthropology, comparative studies of religion, political economy and Marxism have made all this clear.

Yet there seem to be certain constants which all cultures have found 'beautiful': among them – certain flowers, trees, forms of rock, birds, animals, the moon, running water . . .

One is obliged to acknowledge a coincidence or perhaps a congruence. The evolution of natural forms and the evolution of human perception have coincided to produce the phenomenon of a potential recognition: what *is* and what we can see (and by seeing also feel) sometimes meet at a point of affirmation. This point, this coincidence, is two-faced: what has been seen is recognized and affirmed and, at the same time, the seer is affirmed by what he sees. For a brief moment one finds oneself – without the pretensions of a creator – in the position of God in the first chapter of Genesis . . . And he saw that *it was* good. The aesthetic emotion before nature derives, I believe, from this double affirmation.

Yet we do not live in the first chapter of Genesis. We live – if one follows the biblical sequence of events – after the Fall. In any case, we live in a world of suffering in which evil is rampant, a world whose events do not confirm our Being, a world that has to be resisted. It is in this situation that the aesthetic moment offers hope. That we find a crystal or a poppy beautiful means that we are less alone, that we are more deeply inserted into existence than the course of a single life would lead us to believe. I try to describe as accurately as possible the experience in question; my starting point is phenomenological, not deductive; its form, perceived as such, becomes a message that one receives but cannot translate because, in it, all is instantaneous. For an instant, the energy of one's perception becomes inseparable from the energy of the creation.

The aesthetic emotion we feel before a man-made object – such as the white bird with which I started – is a derivative of the emotion we feel before nature. The white bird is an attempt to translate a message received from a real bird. All the languages of art have been developed as an attempt to transform the instantaneous into the permanent. Art supposes that beauty is not an exception – is not *in despite of* – but is the basis for an order.

Several years ago, when considering the historical face of art, I wrote that I judged a work according to whether or not it helped men in the modern world claim their social rights. I hold to that. Art's other, transcendental face raises the question of man's ontological right.

The notion that art is the mirror of nature is one that

only appeals in periods of scepticism. Art does not imitate nature, it imitates a creation, sometimes to propose an alternative world, sometimes simply to amplify, to confirm, to make social the brief hope offered by nature. Art is an organized response to what nature allows us to glimpse occasionally. Art sets out to transform the potential recognition into an unceasing one. It proclaims man in the hope of receiving a surer reply . . . the transcendental face of art is always a form of prayer.

The white wooden bird is wafted by the warm air rising from the stove in the kitchen where the neighbours are drinking. Outside, in minus 25°C, the real birds are freezing to death!

1985

THE WHITENESS OF BIRDS
BY NICHOLAS MIRZOEFF

All we knew was the birds had been debating for ages,
no consensus or conclusion had been reached.
Nathaniel Mackey, "Telling It to the Birds"

The weapon of theory is a conference of the birds.
Fred Moten and Stefano Harney, *All Incomplete*

The just person does not argue for their rights.
It is for others that they stand and fight.
Attar, *The Conference of the Birds*

What is a bird? For the planter and colonist, it was often a pest, eating seed or fruit. For the poor and enslaved, it was a significant source of food, made into a commodity by mass killing. In settler colonial practice, the bird was rendered into a viewpoint, the bird's-eye view that has now become fully automated and digitized. White seeing takes place from the bird's-eye view, conceptually and physically, whether in racial hierarchy or from a balloon, plane, helicopter, or drone. Such "racializing surveillance," to use Simone Browne's term,[1] is now the predicate to any possible racial capitalism.[2] For those whom racializing

---

1 Simone Browne, *Dark Matters: On the Surveillance of Blackness* (Duke University Press, 2015), 15.

2 Cedric J. Robinson, *Black Marxism: The Making of the Black Radical Tradition* (University of North Carolina Press, 2000)

surveillance would contain and segregate, the bird was a visible example of freedom. Bird-watching, by contrast, is a metonymy of settler colonialism. The settler sees the bird, kills it, classifies it, and has it stuffed. Alive, the bird embodies freedom. Dead, first it was an extractive commodity; later, when displayed as an attraction in museums of natural history, it contained and expressed "higher" values of aesthetics. There are over 750 natural history museums in the United States, not to mention 2,400 zoos. Rendered into a commonplace extractive item of exchange, birds index the intersection of extinction, settler colonialism, and racializing capitalism. The removal of birds from colonized land clears the air and makes it open for militarized surveillance. In Palestine, the Israeli Defense Force continues to de-bird the occupied territories. For when the ground is claimed by settlers as nothing, terra nullius, requiring in response what Sarah Elizabeth Lewis calls "groundwork,"[3] the air still remains. Christina Sharpe evokes the "air of freedom," no metaphor to those incarcerated, and the possibility of aspiration.[4] In the midst of the pandemic of a respiratory disease and one year in the wake of George Floyd, on the same day as the Central Park bird-watching incident, these questions are, again, in the air.

---

3 Sarah Elizabeth Lewis, "Groundwork: Race and Aesthetics in the Era of Stand Your Ground Law," Art Journal 79, no. 4 (2020): 92–113.

4 Christina Sharpe, *In the Wake: On Blackness and Being* (Duke University Press, 2016), 104–12; Nicole Fleetwood, *Marking Time: Art in the Age of Mass Incarceration* (Harvard University Press, 2020), 34–35, 52–53.

EXTRACTION

One of the foundational mythologies of settler colonialism has been white people's claim to dominion over all the flora and fauna of the planet as an inexhaustible resource. Moving beyond even the divine dominion given to Adam in Genesis, John Locke's *Second Treatise of Government* claimed: "Subduing or cultivating the earth and having dominion, we see are joined together."[5] Land could be taken, according to Locke, if it was not being cultivated—that is, if it was seen as terra nullius (nothing land), the Roman legal doctrine. Such cultivation of terra nullius gave dominion over land and life alike. Empire was coextensive with white nature—all subdued earth was both empire and white nature. Macarena Gómez-Barris succinctly summarizes: "European colonization throughout the world cast nature as the other and, through the gaze of terra nullius, represented Indigenous peoples as non-existent."[6] Gómez-Barris here insightfully considers terra nullius as a gaze. Someone arrives, looks at land, fails to see who and what is already there, and claims it for themselves. There will be maps, plats, titles, and the other apparatus of bureaucratic colonizing. The result is what she calls "the extractive zone." The extractions are not just minerals, ores, and other raw materials; they are also birds, bison, fossils, archaeological traces, human remains, photographs, and moving images. That extraction is then turned into value, whether financial or cultural.

---

5 John Locke, "Second Treatise of Government" in *Two Treatises on Civil Government* (George Routledge and Sons, 1884), chap. 5, section 35.

6 Macarena Gómez-Barris, *The Extractive Zone: Social Ecologies and Decolonial Perspectives* (Duke University Press, 2017), 6.

After the Haitian Revolution (1791–1801), with the terra nullius having risen up and created its own gaze, European natural history added extinction to extraction as a way of (un)seeing, as Ursula Heise and others have shown.[7] Formerly considered heretical or impossible, extinction—also known as catastrophe or revolution—transformed natural history into life science (biology) in the era of the revolutions of the enslaved and abolition. Based on observation and display, the gaze of extinction, to adapt Gómez-Barris, served as evidence for white superiority by means of purportedly superior capacity for visual observation. Audubon claimed his art depicted "*nature as it existed.*"[8] What geologist, comparative zoologist, and racial theorist Louis Agassiz termed "the naturalist's gaze" included extinction as a past or present possibility. Twinned natural history and anthropological museums were formed in order to collect species and specimens, whether human or other-than-human, before they became extinct. Under the gaze of extinction, those colonized should also be collected. Collection became the third leg of colonial administration, added to the longstanding imperatives of ordering and governing set by Barbados's slave law in 1660.[9] After Haiti, and in the museum, the very idea of "nature" was and is inextricably entangled with race. There are thirty-six thousand African objects in the American Museum of Natural History.

---

7 Ursula Heise, *Imagining Extinction: The Cultural Meaning of Endangered Species* (University of Chicago Press, 2016); and Richard Grusin, ed., "Introduction," in *After Extinction* (University of Minnesota Press, 2018), vii–xx.

8 John James Audubon, *Writings and Drawings* (Library of America, 1999), 754.

9 Tony Bennett, Fiona Cameron, Nélia Dias, et al., *Collecting, Ordering, Governing: Anthropology, Museums, and Liberal Government* (Duke University Press, 2017), 9–51.

In 2021, the Smithsonian Institution revealed that it holds thirty thousand "items" of human remains, while Harvard University has a further twenty thousand. Nearly all were taken from Indigenous peoples, with a sliver devoted to the enslaved. All were said to be required for the gaze of extinction to consider.

In this frame of colonizing and extinction, the ecocide of bird populations is both persistent and pervasive. Birds were slaughtered in totally disproportionate numbers to any need or market. Long offered as commodities in the settler economy, sold for pennies in local markets, birds finally acquired some commercial and noncommercial value by virtue of scarcity on the verge of extinction. But the immense labor of causing extinction was itself a form of extraction. It eliminated a certain possibility of freedom and turned it into a commodity. The ornithologist Audubon recorded that the price of a passenger pigeon in New York went up from a penny in 1805 to four cents in 1830.[10] That's a basic extractive commodity in a period where a laborer made about a dollar a day and farmed poultry was around twelve cents a pound. But, mused Audubon, their numbers did not seem to decrease. By that logic, slaughtering appeared to generate not scarcity but a modest rise in value. It made a certain kind of financial sense to kill everything in sight. But creating endangered species, human and other-than-human, was a more important component of the psychological wages of whiteness. Exterminating birds offered a particular form of violent pleasure by confirming and making visible the capacity of colonization.

---

10 Audubon, *Writings*, 267.

MAY 25, 2020

May 25, 2020, the day on which George Floyd was murdered by Derek Chauvin in Minneapolis, began with what became known as the Central Park bird-watching incident. It demonstrated that it is still inconceivable for a white woman with a University of Chicago master's degree and a career in finance to conceptualize bird-watching while Black. As types, blackness and birds, in the worldview personified by the dog walker Amy Cooper, are the objects of white taxonomy, not its performing subjects. New York City Audubon Society board member Christian Cooper asked her to leash her dog, as required to protect the wildlife. Instead she saw him not as a person but as a type, "African American," as if she was observing wildlife. Her racialized seeing transformed his spoken request into a violent assault. This form of looking was known under Jim Crow as "reckless eyeballing," meaning to look a whiter person in the eye, especially as a form of (alleged) sexual desire. Unable to envisage that a Black man might instead be looking at birds, Amy Cooper called the NYPD. By the time they arrived, Christian Cooper had left, but he released the cell-phone video he made of the incident. Amy Cooper's racialized seeing in Central Park was not an exception: it was constitutive of her role in finance capital. She worked for a trifecta of 2008 financial crash companies: Lehman Brothers, AIG Insurance, and Citigroup. While many others were ruined, Cooper did just fine, ending up as a head of insurance portfolio management at Franklin Templeton, a hedge fund managing $1.5 trillion in assets. Its webpage devoted to diversity—"top to bottom"—then showed a white woman like Cooper. Police in Minneapolis would view George Floyd in related racializing terms. It was only seventeen-year-old Darnella Frazier's cell-phone video that prevented Floyd from becoming just another statistic. When the prosecutor invited the jury to "believe what you saw," it marked one of the very few occasions when such evidence was indeed taken at face value. By the same token, as much as it might be comforting to think of Amy Cooper as an exception, she was more exactly an example of the white way of seeing that renders life into types within a racializing hierarchy. And financializes the result. Astonishingly, exactly one year later, Cooper sued investment firm Franklin Templeton, her former employers, for wrongful dismissal on the grounds of racial discrimination. Cooper claims to have been fired *because* she is a white woman, exemplifying the resurgence of white fragility as a claim to white supremacy.[11]

Meanwhile, the Audubon Society had warned in 2019 of "a net loss approaching 3 billion birds, or 29% of 1970 abundance"—following from decades of such reports, stretching back to the nineteenth century. What birds there are to see now are a fraction of a fraction of a fraction of what there used to be. After 125 years of warnings, the de-birding of the settler colony is close to complete. The remaining sliver of avian life is sustained by human feeding, reserves, zoos, and the like as an attraction, as entertainment, and as a hobby. This is not an accident. The active extermination of Indigeneity was the product of the same elements that comprised the Central Park incident: extraction, whiteness, other-than-white life, and natural history. The result is not the expected white paradise but what the scientists call "an ecosystem collapse."[12]

---

11 Jonah Bromwich and Ed Shanahan, "Amy Cooper, White Woman Who Called 911 on Black Birder, Sues over Firing," *New York Times*, May 26, 2021.

12 Gustave Axelson, "Nearly 30% of Birds in the US, Canada Have Vanished since 1970," *Cornell Chronicle*, September 19, 2019.

FUGITIVITY

In the United States, there has been a particular relation between Atlantic slavery and birds that can only be outlined here. If, for the colonist, birds were an index of whiteness as property within terra nullius, they were for the enslaved the sight of what Walter Johnson calls "freedom as a bodily practice." The enslaved dreamed of flight, and envied the birds their movement, while slaveowners feared that birds might convey ideas and practices of resistance.[13] Even bird-watching and ornithology in the United States cannot be separated from slavery. The Saint-Domingue born slave-owner turned naturalist Jean-Jacques Audubon produced his legendary *Birds of America* as a taxonomy of terra nullius (or perhaps *aer nullius*, the nothing air). His name now authorizes the Audubon Society, for whom Christian Cooper serves as a board member in New York. In turn, as a refugee from one plantation economy turned settler in another, Audubon often had to decide who he was, indicated by his long list of names: John James Audubon, Jean-Jacques Rabin, Jean-Jacques La Forêt, and John James La Forest.[14] Born in Saint-Domingue to a plantation owner and enslaver father and a Jewish servant mother, Jeanne Rabin, he became a refugee in France from postindependence Haiti. Finding his way to the United States as an adult, Audubon was haunted by abolition, the extinction of birds, and of the Indigenous population—whom he saw as doomed—and indeed of the American wilderness as such.[15]

---

13 Walter Johnson, *River of Dark Dreams: Slavery and Empire in the Cotton Kingdom* (Belknap Press of Harvard University Press, 2013), 209–210.

14 Richard Rhodes, *John James Audubon: The Making of an American* (Knopf, 2004), 4–5.

15 Audubon, *Writings*, 522.

Audubon turned to writing about birds after his debt-funded purchase of enslaved people to work at his Kentucky mill ended in bankruptcy in 1819. In his 1826 Mississippi River journal, written in his idiosyncratic Franglais, Audubon discovered a new means of accumulation through ornithology: "So Strong is my Anthusiast to Enlarge the Ornithological Knowledge of My Country that I felt as if I wish myself *Rich again*."[16] His last human property rowed him down the Mississippi to New Orleans, where he sold the two men. His famous drawings of birds were less original than many assume. Their large format and "action" poses were standard at the time in French (if not North American) ornithology.[17] His scenes were not drawn from life in the wild but made in his studio, using dead birds suspended by wires. He worked in a variety of media but not oil paint. One of his most influential drawings was that of what he called the wild pigeon, now usually known as the passenger pigeon, because it became extinct in August 1914. Audubon's drawing stands in for the absent birds. The unreal sharpness of line; the sense that the body was assembled in geometric sections, which is to say, the body as a machine; and the hyperreal clarity of the colonized white space used as the ground combine to give a paradoxical sense of both precision and abstraction *(fig. 1)*. In a word: whiteness.

---

16 Audubon, *Writings*, 47.

17 Linda Dugan Partridge, "By the Book: Audubon and the Tradition of Ornithological Illustration," *Huntington Library Quarterly* 59, nos. 2/3 (1996): 269–301.

FIG. 1 John James Audubon, *Passenger Pigeon* (1828–35). University of Pittsburgh via Wikimedia Commons. Public domain.

Audubon's doubled settler status enabled his seeing of birds as colonial accumulation and as part of the property interest of whiteness.[18] He authorized this technique as derived from the studio—a combination of art school and picture workshop—of the great neoclassical painter Jacques-Louis David.[19] While Audubon did have artistic training in France, there is no record of his having been part of David's extensive studio. David's style certainly centered on the depiction of line to create form, just as Audubon's work did. In painting of the period, to use "line" to structure the image, rather than "color," was to define your work as History, meaning its most serious and morally important category. What History was Audubon depicting? It was the acceleration of the de-birding and de-wilding of America as an index of the formation of the settler colonial United States after the Louisiana Purchase. His originality was to express the resulting tensions between race, colonization, and extinction in a non-human but evocative form of History, which is to say, birds.

Fugitivity was a repeated figure in Audubon's work. In his *Ornithological Biography*, published as a textual accompaniment to the famous pictures, Audubon claimed to have encountered a maroon[20] (whom he called a runaway slave) in the Louisiana bayou, living in a canebrake with his family.[21] Here was yet another *homme de la forêt*,

---

18 Cheryl Harris, "Whiteness as Property," *Harvard Law Review* 106, no. 8 (1993): 1710–93.

19 John James Audubon, *Prospectus: Birds of America* (Adam Black, 1831), 8.

20 Sylviane A. Diouf, *Slavery's Exiles: The Story of the American Maroons* (New York University Press, 2016), 87.

21 John James Audubon, "The Runaway," in *Ornithological Biography: or, An Account of the Habits of the Birds of the United States of America*, vol. 2. (Adam and Charles Black, 1843), 27–32.

or as Audubon put it, mixing racialized metaphors: "a perfect Indian in his knowledge of the woods." Developing his story, Audubon tells how this man had been resold following the bankruptcy of his first owner, separating his family. He memorized the destination of his wife and children, and after he himself had escaped, rescued them, and, with the cooperation of those still enslaved, made camp in the woods. The bankruptcy and family breakup again echo Audubon's personal, rather than ornithological, biography. He devised a fantasy ending in which the maroons obeyed him because of their "long habit of submission" and returned with him to their original plantation, where Audubon persuaded the new owner to take them all into his ownership. He ends his little reverie with the inaccurate statement that since this time it has "become illegal to separate slave families without their consent." The pursuit and biography of birds led Audubon to imagine personal and political reconciliation within racial hierarchy and restored slavery, as if the Haitian Revolution had never happened. For him, the restoration of benevolent slavery was a happy ending. There was no illustration for "The Runaway."

By 1851 a white doctor, Samuel Cartwright, claimed to identify "drapetomania," or the "mental alienation" of running away among the enslaved. Reporting in the year of the Fugitive Slave Act to the Medical Association of Louisiana "on the diseases and physical peculiarities of the Negro race," Cartwright claimed to be a pioneer of "observation," the Audubon of enslaved human beings.[22] For Cartwright the evidence for African incapacity was the absence of any progress in arts and sciences, including

---

22 Samuel A. Cartwright, "Report on the Diseases and Physical Peculiarities of the Negro Race," *New Orleans Medical and Surgical Journal* (May 1851): 691–715.

monuments. Inevitably, he added that if left alone they "would relapse into barbarism, or into slavery, as they have done in Hayti."[23] Based on these observations, Cartwright named and categorized drapetomania. Cartwright further diagnosed a wide-ranging "Dysaesthesia Aethiopica" (Black dysaesthesia), a "hebetude," or laziness, of mind and body, especially prevalent among "free negroes." Dysaesthesia was the opposite of the properly aesthetic conquest of nature, which he might have called euaesthesia (good aesthetic). Freedom was dysaesthetic, an alienation. Dysaesthesia caused the enslaved to "break, waste and destroy everything they handle. […] They wander about at night. […] They slight their work. […] They raise disturbances with their overseer." All of these actions were obviously modes of refusal and resistance. Noting how the enslaved destroyed the plants they were supposed to cultivate, broke their tools, tore their clothes, took things, and refused to respond to punishment, Cartwright inadvertently described the general strike against slavery in the making. Cartwright instead diagnosed them to be symptoms of lung disease, caused by "blood not sufficiently vitalized being distributed to the brain." Similar beliefs have found their way into present-day police reports to account for the death of Black suspects by chokehold. For Cartwright, it was only slavery that beneficially produced the necessary "exercise" to "decarbonize their blood."[24]

---

23 Cartwright, "Report on the Diseases," 694.
24 Cartwright, "Report on the Diseases," 711–13.

"NECROGRAPHY"

Audubon drew birds in the frame of a naturalized plantation economy, a "good aesthetic."[25] In 1831 Audubon observed what he called the snowy heron, also known as the white egret, near Charleston in South Carolina. In the background, Audubon's assistants had painted a plantation called Rice Hope, where the enslaved cultivated rice. None are to be seen here. Instead, Audubon, masked as bird hunters tend to be, is seen with his rifle, chasing fugitives, whether birds or people. The print represented supremacy as the intersection of whiteness, settler colonialism, the Second Amendment, and the invisibility of enslaved African labor. You can buy originals and reproductions of it all over the internet, teaching racialized vision, one print at a time. Audubon casually recorded how practices of enslavement affected even common songbirds. For instance, the blue jay was a prolific species with a habit of eating crops, so that in Louisiana "the planters are in the habit of occasionally soaking some corn in a solution of arsenic, and scattering the seeds over the ground, in consequence of which many Jays are found dead about the fields and gardens."[26] He did not need to mention that in Louisiana all planters used enslaved labor *(fig. 2)*.

---

25 I borrow the term "necrography" from Dan Hicks, *The Brutish Museums: The Benin Bronzes, Colonial Violence and Cultural Restitution* (Pluto, 2020), 25–37.

26 Audubon, *Writings*, 291.

FIG. 2 John James Audubon, *Snowy Heron* (1828–35). University of Pittsburgh via Wikimedia Commons. Public domain.

Audubon's writing was no pastoral. Rather, it was an account of how much killing is involved in making a settler colony. And making pictures from the results. A single blast from a shotgun killed 120 blue-winged teals, a duck, in New Orleans.[27] Audubon saw one man in Pennsylvania kill 6,000 passenger pigeons in a single day. While the passenger pigeon has become a somewhat notorious example, almost every bird Audubon looked at comes with a story of mass slaughter. Take the golden plover. While in Louisiana, Audubon noted: "The gunners had assembled in parties of from twenty to fifty at different places, where they knew from experience that the Plovers would pass." He estimated that 144,000 were killed. As a result, "the next morning the markets were amply supplied with Plovers" at a very low price.[28] Cannier hunters would have brought fewer birds and made more. But by removing birds from the wild, hunters deprived both the enslaved and Indigenous of a food resource and forced the settler poor to buy them rather than hunt them.[29] For bird hunters, killing was part of the emotional wages of whiteness, where to be "master" as Audubon had been in the bayou, whether over nature or the enslaved, was itself the reward. This is exactly what a certain kind of white person still says: give me liberty or give me death, where liberty is defined by access to guns.

By the late nineteenth century, these massacres had gone so far that new colonial institutions were created to preserve remnants as specimens and to display the past

---

27 Audubon, *Writings*, 476.

28 Audubon, *Writings*, 87–88.

29 For example, Audubon described how Africans in Louisiana would make "gombo soup" from brown pelicans, of which "they kill all they can find" (*Writings*, 455).

in museums. The real problem with the American Museum of Natural History is not its displays—as awful as they are—but the fact that it had to be created at all. The institutionalization of "nature" through museums, zoos, wildlife refuges, national parks, and the definition of wilderness marked the peak of empire's exterminations. By consigning the other-than-human world to this status as an attraction to be looked at, empire claimed full dominion over life. The dead animals displayed in the dioramas of the American Museum of Natural History teach children both that the role of non-human life is to die and that they are to be its killers. Its opening in 1877 was the corollary to the creation of anthropological museums in Europe, formed from the loot taken by colonial expeditions. Just as Dan Hicks sees the anthropological museum as an "implement of ... imperialism made in the final third of the 19th century,"[30] so too were museums of natural history, game reserves, and national parks tools of enshrining settler colonial dominion over all nature and making it permanent. This maneuver further closed the distinction between enclosed/colonized space and environment/uncolonized space, with the dangerous exception of the so-called reservation. As one part of this far-reaching process, ornithology was separated from natural history and institutionalized with the formation of the American Ornithological Union in 1883.[31] Its function was to assert the primacy of other-than-exchange value for birds and, by extension, the natural world as a whole.

---

30 Hicks, *The Brutish Museums*, 9.

31 Dorceta E. Taylor, "Blaming Women, Immigrants, and Minorities for Bird Destruction," in *The Rise of the American Conservation Movement: Power, Privilege, and Environmental Protection* (Duke University Press, 2016), 190–223.

EUGENIC DIORAMA

In 1886 a special supplement of the new magazine *Science* was devoted to the decline of bird species.[32] American Museum of Natural History curator J. A. Allen exhorted his readers that despite their still-limited financial value, "birds may be said to have a practical value of high importance and an aesthetic value not easily overestimated." He blamed five groups for their decline: market gunners, who hunted game birds; African Americans who trapped and sold small birds across the South for food; women, who wore hats with feathers, named as "the dead bird wearing gender"; immigrants who trapped birds for food; and even small boys who hunted for eggs. From this optic, the survival of birds was a blueprint for the eugenic survival of white supremacy in the uncertain conditions of modernity, which meant "reconfiguring the terms of being natural in culture."[33] The eugenic program developed by Francis Galton and enthusiastically received in the United States offered solutions to all the issues highlighted by Allen.[34] White women were to give up their monstrous display of feathers and return to breeding the "great race." Immigration was to cease, as the 1882 Asian Exclusion Act prefigured, followed by ever more restrictive legislation. Jim Crow would reassert control of the South, and public schools would discipline children. The United States found many other ways for

---

32 J. A. Allen, "The Present Wholesale Destruction of Bird-Life in the United States," *Science* 7, no. 160 (1886): 191–95.

33 Emily Gephart and Michael Ross, eds., "How to Wear the Feather: Bird Hats and Ecocritical Aesthetics," in *Ecocriticism and the Anthropocene in Nineteenth-Century Art and Visual Culture* (Taylor and Francis, 2019), 192–207.

34 Daniel J. Kevles, *In the Name of Eugenics: Genetics and the Uses of Human Heredity* (Harvard University Press, 1995).

men to satisfy their desire to shoot. But none of these eugenic policies much helped the birds. Americans continued to eat all kinds of bird in the period, including meadowlarks, blackbirds, sparrows, thrushes, warblers, vireos, waxwings, reed-birds, robins, and flickers. In 1897 William Hornaday of the New York Zoological Society concluded from two hundred questionnaires sent around the country that nearly half of all American birds had perished since 1882. As late as 1940 Eleanor Roosevelt wrote a column asking women to cut back on wearing feathers.

The display of animals in habitat groups within dioramas in natural history museums came to the United States in 1887 with the display of eighteen groups of nesting birds with accessories like foliage and flowers, created by the British museum modeler Mrs. E. S. Mogridge (d. 1903), who had worked for the British Museum and its Museum of Natural History with her brother H. Mintorn.[35] While using just a fraction of the American Museum of Natural History's forty thousand birds, Mogridge's models changed its practice dramatically. She was credited by museum director Henry Fairfield Osborn for inspiring its famous habitat displays.[36] Beginning in 1902 the president of the American Ornithologists Union, Frank Chapman, created a group of bird dioramas at the American Museum of Natural History.[37] In Allen's contemporary description: "The area of these groups ranges from 60 to

---

35 "Notes and News," *The Auk* 20, no. 3 (1903): 326–30. Her first names are not reported.

36 Henry Fairfield Osborn, *Creative Education in School, College, University, and Museum: Personal Observation and Experience of the Half-Century 1877–1927* (Charles Scribner's Sons, 1927), 235.

37 Elizabeth Barlow Rogers, "Representing Nature: The Dioramas of the American Museum of Natural History," *SiteLINES: A Journal of Place* 8, no. 2 (2013): 10–14.

160 square feet, to which is added a panoramic background, which in most cases merges insensibly into the group itself. The backgrounds are painted by skillful artists, generally from studies made at the actual site represented."[38] The panorama depicting wading birds was based on Cuthbert Rookery in Florida, where a game warden had been killed by feather poachers in 1905 and the birds were believed to be at risk of local extinction. Chapman conceived of the panoramas as three-dimensional perspective paintings, noting "the background is curved [convex backward] with the front opening so reduced in size that at the proper distance, or 'correct view-point,' neither the ends nor the top of the group can be seen. By thus leaving the actual limits of the group to the imagination the illusion of space and distance is greatly heightened."[39] The ideal spectator would then stand at the proper viewpoint and be taken in by the illusion, to get a bird's-eye view of the birds *(fig. 3)*.

38 J. A. Allen, "The Habitat Groups of North American Birds in the American Museum of Natural History," *The Auk* 26, no. 2 (1909): 165–74.

39 Quoted in Allen, "Habitat Groups," 174.

FIG. 3 Wading Bird diorama at the American Museum of Natural History, 1909.

THE ASSEMBLY OF BIRDS

In a 2021 interview, Fred Moten and Stefano Harney stress how a key phrase to understand *All Incomplete* is the "conference of birds" and the murmurations that result. The word *murmuration* describes a flock of birds moving together to create fractal patterns in the sky. These moments are mesmerizing. Go and look at one of these videos.[40] Afterward, imagine a murmuration with millions of birds in it, the flocks of birds that were commonplace before settler colonists systematically killed them all. Audubon described the "angles, curves and undulations" in the "multitudes of Wild Pigeons" as tactics designed to repel hawks.[41] He estimated that there were no less than a billion birds in the murmuration. Now you have a sense of what the de-birding of North America has wrought. In his account of slavery, Frederick Law Olmsted, one of the architects of Central Park, often used the term *murmur* to mean conspire or revolt.[42] A murmuration in that sense is the revolution. Perhaps the conference of birds is the past future that awaits.

In seventeenth-century Europe, the "assembly of birds" was a popular artistic theme, often treated by the Dutch artist Frans Snyders. The parrot and toucan in one version make it clear that it was always a colonial imaginary. The owl at the center is either directing the song or warning the birds of the human hunters following Aesop's fable. Or both. I like to think of it as an actual assembly, a democracy in which predators and songbirds have an

40 Marco Valk, "Starling Murmuration, or The Dance of the Starlings," YouTube video, 5:02, posted February 8, 2020.

41 Audubon, *Writings*, 263, 262.

42 Frederick Law Olmsted, *A Journey in the Seaboard Slave States* (Dix and Edwards, 1856), 191–94.

equal say. Unlike humans, except in opera, there's no need for them to speak one at a time. Their chorus forms what settlers had heard as the "wild jubilee," a celebration of the end of slavery in nature. There's a *Concert of Birds* (ca. 1630) by Snyders in the Prado without the owl, which is to say, an assembly without a visible leader, a direct democracy of other-than-human life *(fig. 4)*. Some suggest that the owl was cut out at some point, but no matter. Someone wanted it to look like that.

FIG. 4 Frans Snyders, *Concert of Birds* (1630), Museo del Prado, Madrid. Photograph via Wikimedia Commons.

Such assemblies have existed and do exist. In the remarkable paintings of Aotearoa, New Zealand artist Bill Hammond, who sadly passed in February 2021, zoomorphic birds dominate the land. Hammond's work radically changed when he took a visit to Motu Maha, or the Auckland Islands, an archipelago to the south of Aotearoa that is designated as an "area outside territorial authority." Early Polynesian settlement gave way to a whaling and sealing station until the immense massacre of the seals made the labor of extermination unprofitable around 1894. The islands are now a national nature reserve, and the authorities are attempting to remove all introduced mammals, like feral cats, rabbits, and pigs. This effort is what Juno Salazar Parreñas calls "decolonizing extinction. Care is not necessarily affection, but for me it is a concern about the treatment and welfare of others."[43] It might mean rehabilitating animals to the wild, as Parreñas describes in the case of orangutans born in captivity and raised "semi-wild." Or it might mean rehabilitating the wild to animals, as is taking place in Motu Maha. On his visit, Hammond was struck by the way the birds on the island stood upright staring out to sea, and it reminded him that prior to the arrival of humans around 1300 CE, birds were the dominant species in the islands now known as Aotearoa. There were no predators, other than the pouakai (Haast's eagle), a giant eagle. The most populous species was the large bipedal flightless birds called moa, which were quickly made extinct by the first human colonists, now known as the Māori. They simply ate them all, in about two hundred years.

---

43 Juno Salazar Parreñas, *Decolonizing Extinction* (Duke University Press, 2018), 6.

44 Richard Wolfe, *Moa: The Dramatic Story of the Discovery of a Giant Bird* (Penguin, 2003), 181.

Archeologists have identified three hundred moa-hunting sites in Aotearoa. Remains suggest that twenty thousand to ninety thousand birds were eaten at the Waitaki River bone midden alone.[44]

By 1873 the naturalist William Buller articulated the ornithologist's task in settler colonialism:

> ***Under the changed physical conditions of the country, brought about by the operations of colonization, some of these remarkable forms have already become almost, if not quite, extinct, and others are fast expiring. It has been the author's desire to collect and place on record a complete history of these birds before their final extirpation shall have rendered such a task impossible; and it will be his aim to produce a book at once acceptable to scientific men in general and useful to his fellow colonists.***[45]

The conditions were different indeed. The Māoris, he noted, said of the kākāpō, or owl parrot (*Stringops habroptilus*), "In winter they assemble in large numbers, as if for business; for after confabulating together for some time with great uproar, they march off in bands in different directions."[46] Buller recognized the sound of this freedom in the dawn chorus, comprising bell-birds, tui, whiteheads, and piopio: "Shortly after daylight a number of birds of various kinds join their voices in a wild jubilee of song."[47] Such natural democracy was, he believed, certain to become extinct as the deliberate result of the operations

---

45 Walter Lowry Buller, *A History of the Birds of New Zealand* (John van Voorst, 1873), 2.

46 Buller, *Birds of New Zealand*, 34.

47 Buller, *Birds of New Zealand*, 136.

of colonization. Indeed, some thirty-five species that he might have heard have since become extinct.

Nonetheless, Māori proverbs recorded the speech of the assembled birds:

E koekoe te tūī, e ketekete te kākā,
e kūkū te kereru

The tui chatters, the parrot gabbles,
the wood pigeon coos.

The speech of the birds enters human language here, their sounds forming words, just as murmuring melds into thinking. When the Europeans came and began doing what colonizers do, a new saying arose among those who, for the first time were now Māori (ordinary), as opposed to members of specific iwi (peoples):

Ka ngaro ā-moa te iwi nei

The people will disappear like the moa.[48]

The Māori may have been the cause of the moa's disappearance, but this saying suggests that they regretted it.

Musing on all this, after reading Buller, Hammond created a massive *Fall of Icarus* (1995), in which the birds watch the human flight of Icarus end in failure. The space is a deep metallic blue, mingled with drips of black and white paint throughout. One of those drips, perhaps the white streaky column in the center-right, marks

48 Quoted and translated by Priscilla Wehi, Hēmi Whaanga, and Murray Cox, "Dead as the Moa: Oral Traditions Show That Early Māori Recognised Extinction," *The Conversation*, September 5, 2018.

Icarus falling from his overambitious flight. Icarus would have been the first (white) colonist in Aotearoa had he been able to land. It might have been a hard place to fly, because there are two distinct horizons, marked by smoking volcanic islands. The birds stand in trees to the left and right from top to bottom, but the space nonetheless "works," visually. The land is new, there are not yet human words to describe it, but the birds sing it into being. What are these two spaces? Colonial and Indigenous? Human and avian? Or a double vision of space, not yet colonized as space, not subject to the "laws" of perspective? Can you fly drones here? The birds look out from the trees, bipedal, winged, armed in the sense that they have arms, long-beaked, wearing one-piece costumes like superheroes. The stillness of their looking watches over this doubled space to live and keeps it possible, where humans would insist otherwise. Some stand on branches, others hang by their "arms." What are the birds actively doing? They create community, they strike against colonialism and the invasion of humans, and they imagine the future. They are waiting, not for Godot, but for the end of whiteness.

THE SILENT ACADEMY
CHAPTER 2: THE LYREBIRD

BY YURI TUMA
Institute for Postnatural Studies

I was supposed to meet Joy at dawn in the forbidden forest, where we planned to practice the bird whistling language we were learning at the secret Silent Academy. The damp ground muffled my footsteps and released a rich, earthy scent with every step. As I ventured deeper, ancient trees spoke to me through the language of the wind and the occasional crack of a trunk. The Lovelessness regime's Protection Act outlawed direct interaction with the other-than-human beings in this nature reserve. My eyes strained in the dim light as I searched for any sign of Joy and feared glimpsing one of the surveillance patrols instead. As I neared our meeting spot for overnight rest, I noticed faint markings etched into the bark of some of the trees—secret symbols of resistance from the Silent Academy.

No one paid much attention when Bowie spoke about the black-noise-technology in that interview with Dick Cavett in 1974:

> **One facet of black noise is that everything—like a glass—[sic] if an opera singer hits a certain note, the vibrations alter the metabolism of the glass and it cracks it. So black noise is the frequency at which you can crack a city or people.**

Unlike white noise, which contains the complete spectrum of all frequencies, black noise is the tangible absence of it, the lack of imperceptible infra-sounds that surround us. It was the 1970s, a time when the concept of human mutualism, let alone intra-species mutualism, was still largely foreign to most. It's unsurprising that a society built on hierarchical systems, particularly the oppressive structures of heteropatriarchy, ultimately gave rise to the Lovelessness regime. Since the failed uprising of the Listening New World Order – the resistance movement against these petromasculine sonic structures – our society has been systematically silenced. First, the regime, driven by its narcissistic hurt, injected humans with a microchip that suppressed our ability to speak, leaving us with only the resonance of inner speech and post-traumatic thoughts. This internal dialogue was a constant reminder of the difficulty of feeling emotions without translating them into spoken words. After the implementation of the microchips, murmurs had circulated about the regime's development of "black noise weaponry." We were warned by infiltrators that those who might survive future detonations would face a peculiar auditory limitation: the ability to hear only birdsong. I scoffed at the notion, dismissing it as propaganda of terror.

Still, since the microchip injections, the Silent Academy charter within the archipelago has been covertly fostering the growth of small communities dedicated to exploring alternative forms of communication, focusing in particular on the language of birds. Their extensive curriculum on practices of active listening and caregiving has provided hope for a future based on love and mutual respect, for living and non-living entities, the human, the more-the-human, and the other-than-human. At the academy we learned to whistle vocalizations consciously, and to attune our pitches and tones to the frequencies of all others, as well as the territory.

We also learned new body language to show affection and relay fear or danger, and we made new head accessories that displayed these among other emotions. The absence of verbal language allowed us to connect on a deeper level. Active listening came as second nature once we were set free from the charged meanings and selfish, subjective interpretations that often accompanied words. By verbally silencing us, the regime thought they could control us, control the regeneration of our communities of resistance. They believed that a third uprising would not happen again, but they were sorely mistaken.

Or so I thought. As I lay awake in the forbidden forest waiting for Joy to arrive, the serene silence of the night was interrupted by a series of deafening shatters and roars of destruction. The black-noise weapons did not explode, but the wreckage of their frequency could be heard loud and clear. The ground trembled and the air was filled with a dissonance of noise that quickly drowned out all other sounds. When the tremors subsided, an eerie silence descended inside my ears, and I realized the gravity of what was happening: the Aural War had begun. Deafened and alone, the chilling precision of the regime's warning struck me. I sprang to my feet in urgency, driven by a desperate need to find the solace of birdsong, longing for the recording of the last song sung by the last of the Kauaʻi ʻōʻō. I wandered through the forest's silent ruins of sonographic shapes, my damaged ears straining futilely for any vestige of sound. Unbidden, thoughts of Joy flooded my mind. Had they escaped the blast in time?

Joy and I had met when we were toddlers, orphans adopted by the Silent Academy. Our connection had been instant and profound—twin flames recognizing each other amidst the chaos of the budding revolution. Hidden within the archipelago's misty coves, the Academy became our sanctuary, our mentors, our chosen family. At first, I struggled to grasp the concept of bird language, my clumsy attempts at whistling drawing amused glances from the others. But as we grew up, I found myself attuning to the subtle chirps and trills in the curiosity for life. Joy, always the quicker learner, mastered the vocalizations before me. Their patient guidance and encouragement were fundamental to my progress and belonging. Together, we explored the nuances of pitch and tone, our connection deepening as we learned to communicate without words. The weight of our shared history gnawed at me as it had never before: our parents, freedom fighters for listening rights, were forever silenced by the Lovelessness regime.

Now, as I move through this soundless, deforested section of land, the absence of Joy's birdsong resonates louder than any shatter ever could. Our parents' courage flows through me, giving me strength to remain calm and to avoid having one of my disabling panic attacks. I breathe deeply, anchoring myself in their love. As the panic subsides, my mind clears, allowing recent memories to surface. Just yesterday, at the Silent Academy, we learned a whistling meditation technique as a way to practice generous listening. Our mentor's words echo in my thoughts: "The first responsibility of any uprising rooted in love's energy is to listen." Giving responsible and bountiful attention, they taught us, was key to any form of communication that could lead to true reparation. We were also instructed in guiding others through this meditation. A spark of hope ignites within me. Perhaps, if I whistle it out loud, I might attract a feathery companion, or a human proficient in bird whistling. With renewed purpose, I prepare to break the quiet, ready to connect with whatever—or whoever—might be out there in this phonologically transformed world.

What follows is an incomplete translated transcript in English of the whistling meditation for you to be able to share with someone.

(Find a comfortable position and close your eyes. As you slowly focus on your breath for a couple of minutes, allow the following words to guide you...)

•••••••••••••••••••••••••••••••••••••••

- Imagine you are lying on the warm grass,
the slanted sunrise gently bathing your face.
- You gaze skyward, watching a flock
of birds soar overhead.
- These are no ordinary birds;
some appear to be hybrid species.
- Each bird in this flock is unique,
the first or the last of their kind.

- They fly in unison, yet each sings a different song.

- How many songs can you hear?
- Can you distinguish the individual voices,
their subtle variations?

- One bird starts to descend towards the grass.
- You feel a wave of empathy rising within you,
a deep sense of heart connection.
- As they land softly beside you,
you recognize a profound feeling of kinship.

- What does this bird look like?
- How does the color of their plumage
reflect the new day's sunlight?
- What color are their eyes?

- The bird begins to sing, a song both ancient and new, a melody of resilience and renewal. Listen deeply to their song. Allow the emotions to flow through you.

- The moment the song ends, you switch consciousness with them.

- You are the bird.

- How does it feel to be the last or the first of your kind?

- You feel the gush of air beneath your wings as you prepare to take flight to meet your flock of extraordinary siblings.

- You join your fellow flying travelers and feel a sense of belonging, a deep soul link to this diverse and interconnected community.

- You are not alone.
- You are part of something greater than yourself.

- How does it feel to belong?

- You look at the grass, now below you, and see that the human you left behind has tears in their eyes.

• •

The forbidden forest absorbs the final notes of my whistle, leaving me enveloped in a profound state of mindfulness. Tears trace warm paths down my cheeks, igniting a renewed sense of hope within me. I stand motionless, my senses heightened, straining to hear the faintest whisper of birdsong.

Time stretches, elastic and uncertain, until finally, a sound pierces the silence – the distant growl of an engine, followed by the harsh screech of a saw biting into wood. Machines? I shouldn't be able to hear those anymore. Their presence here defies logic, yet the sound pulls at me. I follow the noise, each memory of a felled tree a stark reminder of the Lovelessness regime's iron grip on our world. The sounds grow louder, more insistent. But as I near the source, fear and expectation give way to wonder. There, in a small clearing, stands a Menura novaehollandiae – a lyrebird. I had been scared of the very thing I was hoping to find. It wasn't the roar of destruction greeting me, but their intricate, haunting mimicry. They stand there, their feathers shimmering with borrowed sounds, a contradictory echo of the past. Each note of their song is a perfect imitation of the machines, yet somehow transformed. The roar of destruction morphs into a song of resilience, the screech of saws not a cry of demolition, but defiance.

Chapter 1:

Death, Dreams, and Other Realms

(22)

(23)

(24)

(31)

(33)

(30)

(32)

(39)

(37)

(38)

(44)

(26)

(27)

(28)

(29)

(36)

(35)

(42)

BIBLIOTHÈQUE MUNICIPALE DE LYON

(41)

(46)

(47)

Fecit olorinis Ledam recubare sub alis.

(48)

BIRDS' NESTS AND THEIR BUILDERS

(31)

(33)

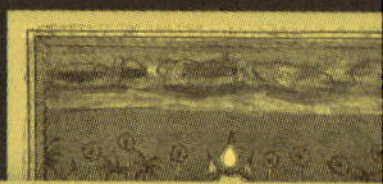

Copyrighted Photograph 1909 by K H Martin
A good day for ducks in Iowa.

(44)

(26)

(27)

(28)

(29)

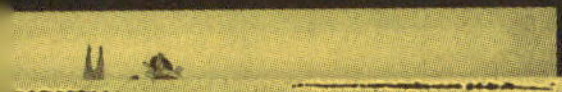

BIBLIOTHEQUE MUNICIPALE DE LYON

(46)

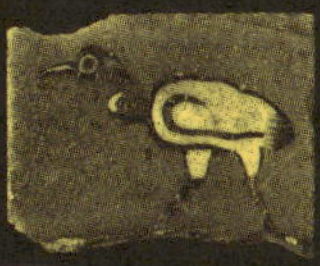

(47)

Fecit olorinis Ledam recubare sub alis.

(48)

(22)

(30)

(31)

(33)

(32)

(37)

(39)

(43)

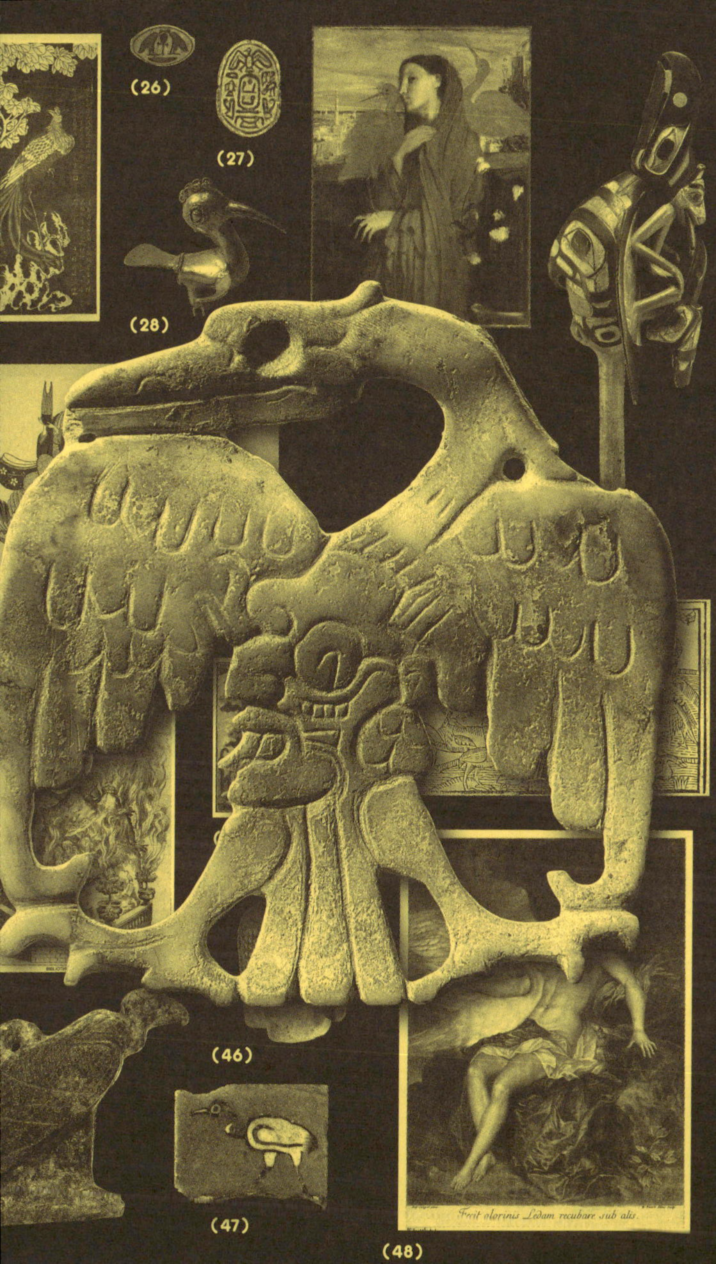
Fecit olorinis Ledam recubare sub alis.

(26)

(27)

(28)

(46)

(47)

(48)

(22)

(39)

)

(43)

BIBLIOTHEQUE MUNICIPALE DE LYON

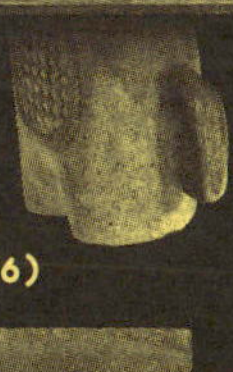

(46)

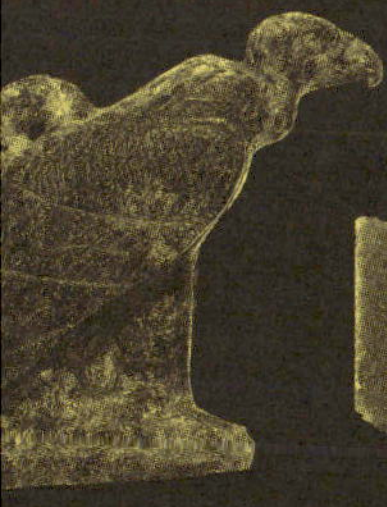

(47)

Fecit olorinis Ledam recubare sub alis.

(48)

(23)
4)
(22)
(30)
ELIJAHFED B
(37)
(39)
(44)
(43)

(26)

(27)

(28)

(29)

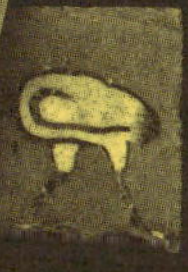

(47)

(48)

(22)

(23)

(24)

(30)

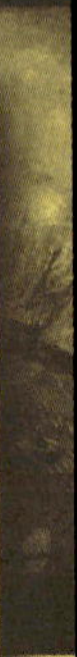

(37)

(38)

(39)

(44)

Fecit olorinis Ledam recubare sub alis.

BELLY FULL OF WORMS
BY MONIKA CZYŻYK

SITE-SPECIFIC PAINTINGS AT ONOMATOPEE
JANUARY 10 – APRIL 27, 2025

Between January 4 and 10, 2025, I created *Belly Full of Worms*, a series of site-specific clay paintings on the glass surfaces of Onomatopee. Due to their materials and placement, these ephemeral paintings disappear once the exhibition ends, leaving only photographic and remembered traces behind.

The birds depicted in clay on the windows are representations of birds I have encountered myself, whether in reality or in dreams. The works emerged from my ongoing exploration of impermanence, mythology, and natural cycles of decay and renewal. The composition and thematic focus of these pieces were guided by the way that birds—in reality and in stories—move between earth and air elements, between life and death, between presence and disappearance.

My work materializes through a process-based method that combines intuition with practical techniques for building up the imagery. Each work reveals itself as I paint, in relation to this process and to image research. The composition for the painting on the entrance windows was influenced by *The Rohan Hours*, a 15th-century illuminated manuscript that I found by chance in an old drawer of abandoned books at the Vartiosaari Artists' House, Finland. While I flipped through this book,[1] its narrow vertical format offered me a rhythm that helped resolve the layout of the rectangular glass surfaces.

One particular image from the manuscript stayed with me: two levitating figures, frozen in agony, hovering above trees filled with birds. Their suspended stillness, in stark contrast with the implied movement of the birds below, encouraged my eyes to train over the image vertically, rather than horizontally as we often would. This verticality indicates not only a physical direction but a position in society, a gaining of power, a journey to a spiritual realm. The visual tension in the image shaped my own window compositions, in which I layered birds, symbols such as ladders and wings, and fragmented figures that shift between presence and dissolution, in different vertical combinations.

Beneath the glass door, I painted the image of a squirrel, its posture eerily peaceful, as though caught between sleep and death. Surrounded by birds, the squirrel represents a moment of opportunity for other neighboring creatures; decomposers. Inspired by Bernd Heinrich's *Life Everlasting*,[2] I was drawn to how birds—vultures, crows, ravens—play a vital role in the natural cycle of decomposition and rebirth.

Each figure in the paintings carries layered meanings and connotations. Two vultures I originally observed in Bulgaria's Rhodope Mountains embody the presence of death, while storks I encountered in a Polish bird sanctuary in

---

1 As a Manifesting Generator 3/5 Heretic Martyr.

2 Bernd Heinrich, *Life Everlasting: The Animal Way of Death* (Mariner Books, 2013).

Złota symbolize renewal. Each year, the stork couple remain at the shelter, while their offspring leave the nest to migrate. A golden chapel perched on owl feet borrows from the symbolism of folklore, where the familiar and the mythical intertwine. Nearby, a chicken cradles a golden egg, and an owl fiercely guards its young from two approaching ravens. These images, drawn from observation and memory, explore themes of protection, transformation, and survival.

I collect the clay I work with from very specific locations, and each time, the memory of my collecting experience is imbued into the material alongside its own memories of place and origin. Golden clay is my favorite material to work with. For these paintings, I collected the clay in Seltún, Iceland, with permission from the local community. Clay formation is a long process, and our bodies often carry similar minerals. As organically forming matter, clay reflects the idea of renewal, while the material's earthy texture and transient quality, directly applied to windows, signifies the process of decay.

As the week-long installation period neared completion, I painted a czyċyk (siskin)—the bird that inspired my surname—on the central double doors of the exhibition space. I extended the golden pigment onto the back wall outside the gallery, creating a circular form. From a particular vantage point, this shape aligned perfectly with the czyċyk's head, giving it an almost otherworldly glow.

The last two paintings were created particularly instinctively, using local clay collected from the waters of the Rhine and Maas rivers, in monochrome tones. Beyond the window stood a fig tree on which birds would occasionally land, merging the living world with the painted visions.

I close with a story, passed on through a friend, that has lingered with me throughout this process:

> *Long before the land formed, the Little Lark and Father Lark sang about the universe. When Father Lark died, there was no earth for burial. So, the Little Lark made a space in the back of his head and buried his father there. And that is how the first memory was created.*

Like the lark carrying memory, these paintings hold fleeting traces—briefly seen, then gone, yet still present in the echoes they leave behind.

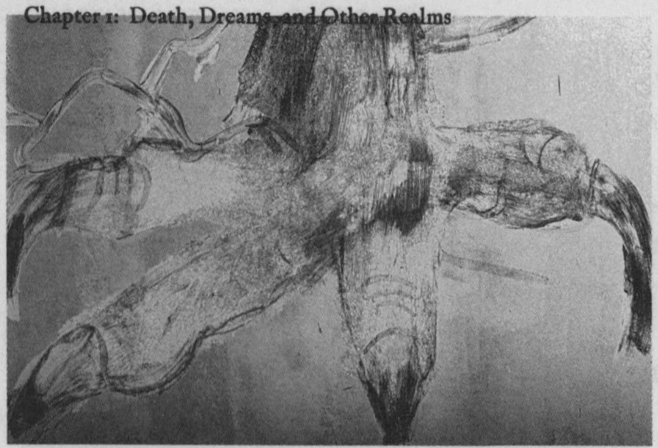

DO BIRDS DREAM?

What new research on the avian brain and REM sleep in birds might reveal about our own dream lives

BY MARIA POPOVA
March 26, 2024

I once dreamed a kiss that hadn't yet happened. I dreamed the angle at which our heads tilted, the fit of my fingers behind her ear, the exact pressure exerted on the lips by this transfer of trust and tenderness.

Freud, who catalyzed the study of dreams with his foundational 1899 treatise, would have discounted this as a mere chimera of the wishful unconscious. But what we have since discovered about the mind—particularly about the dream-rich sleep state of rapid-eye movement, or REM, unknown in Freud's day—suggests another possibility for the adaptive function of these parallel lives in the night.

One cold morning not long after the kiss dream, I watched a young night heron sleep on a naked branch over the pond in Brooklyn Bridge Park, head folded into chest, and found myself wondering whether birds dream.

The recognition that non-human animals dream dates at least as far back as the days of Aristotle, who watched a sleeping dog bark and deemed it unambiguous evidence of mental life. But by the time Descartes catalyzed the

Enlightenment in the 17th century, he had reduced other animals to mere automatons, tainting centuries of science with the assumption that anything unlike us is inherently inferior.

In the 19th century, when the German naturalist Ludwig Edinger performed the first anatomical studies of the bird brain and discovered the absence of a neocortex—the more evolutionarily nascent outer layer of the brain, responsible for complex cognition and creative problem-solving—he dismissed birds as little more than Cartesian puppets of reflex. This view was reinforced in the 20th century by the deviation, led by B. F. Skinner and his pigeons, into behaviorism—a school of thought that considered behavior a Rube Goldberg machine of stimulus and response governed by reflex, disregarding interior mental states and emotional response.

In 1861, just two years after Darwin's publication of *On the Origin of Species*, a fossil was discovered in Germany with the tail and jaws of a reptile and the wings and wishbone of a bird, sparking the revelation that birds had evolved from dinosaurs. We have since learned that, although birds and humans haven't shared a common ancestor in more than 300 million years, a bird's brain is much more similar to ours than to a reptile's. The neuron density of its forebrain—the region engaged with planning, sensory processing, and emotional responses, and on which REM sleep is largely dependent—is comparable to that of primates. At the cellular level, a songbird's brain has a structure, the dorsal ventricular ridge, similar to the mammalian neocortex in function if not shape. (In pigeons and barn owls, the DVR is structured like the human neocortex, with both horizontal and vertical neural circuitry.)

Still, avian brains are also profoundly other, capable of feats unimaginable to us, especially during sleep: many birds sleep with one eye open, even during flight. Migrating species that traverse immense distances at night, like the bar-tailed godwit, which covers the 7,000 miles between Alaska and New Zealand in eight days of continuous flight, engage in unihemispheric sleep, blurring the line between our standard categories of sleep and wakefulness.

But while sleep is an outwardly observable physical behavior, dreaming is an invisible interior experience as mysterious as love—a mystery to which science has brought brain imaging technology to illuminate the inner landscape of the sleeping bird's mind.

The first electroencephalogram of electrical activity in the human brain was recorded in 1924, but EEG was not applied to the study of avian sleep until the 21st century, aided by the even more nascent functional magnetic resonance imaging, developed in the 1990s. The two technologies complement each other. In recording the electrical activity of large populations of neurons near the cortical surface, EEG tracks what neurons do more directly. But fM.R.I. can pinpoint the location of brain activity more precisely through oxygen levels in the blood. Scientists have used these technologies together to study the firing patterns of cells during REM sleep in an effort to deduce the content of dreams.

A study of zebra finches—songbirds whose repertoire is learned, not hard-wired—mapped particular notes of melodies sung in the daytime to neurons firing in the forebrain. Then, during REM, the neurons fired in a similar order: the birds appeared to be rehearsing the songs in their dreams.

An fM.R.I. study of pigeons found that brain regions tasked with visual processing and spatial navigation were active during REM, as were regions responsible for wing action, even though the birds were stilled with sleep: they appeared to be dreaming of flying. The amygdala—a cluster of nuclei responsible for emotional regulation—was also active during REM, hinting at dreams laced with feeling. My night heron was probably dreaming, too—the folded neck is a classic marker of atonia, the loss of muscle tone characteristic of the REM state.

But the most haunting intimation of the research on avian sleep is that without the dreams of birds, we too might be dreamless. No heron, no kiss.

There are two primary groups of living birds: the flightless Palaeognathae, including the ostrich and the kiwi, which have retained certain ancestral reptilian traits, and Neognathae, comprising all other birds. EEG studies of sleeping ostriches have found REM-like activity in the brainstem—a more ancient part of the brain—while in modern birds, as in mammals, this REM-like activity takes place primarily in the more recently developed forebrain.

Several studies of sleeping monotremes—egg-laying mammals like the platypus and the echidna, the evolutionary link between us and birds—also reveal REM-like activity in the brainstem, suggesting that this was the ancestral crucible of REM before it slowly migrated toward the forebrain. If so, the bird brain might be where evolution designed dreams—that secret chamber adjacent to our waking consciousness where we continue to work on the problems that occupy our days. Dmitri Mendeleev, after puzzling long and hard over the arrangement of atomic weights in his waking state, arrived at his periodic table in a dream. "All the elements fell into place as required," he recounted in his diary. "Awakening, I immediately wrote it down on a piece of paper." Stephon Alexander, a cosmologist now at Brown University, dreamed his way to a groundbreaking insight about the role of symmetry in cosmic inflation that earned him a national award from the American Physics Society. For Einstein, the central revelation of relativity took shape in a dream of cows simultaneously jumping up and moving in wavelike motion.

As with the mind, so with the body. Studies have shown that people learning new motor tasks "practice" them in sleep, then perform better while awake. This line of research has also shown how mental visualization helps athletes improve performance. Renata Adler touches on this in her novel, *Speedboat*: "That was a dream," she writes, "but many of the most important things, I find, are the ones learned in your sleep. Speech, tennis, music, skiing, manners, love—you try them waking and perhaps balk at the jump, and then you're over. You've caught the rhythm of them once and for all, in your sleep at night."

It may be that in REM, this gloaming between waking consciousness and the unconscious, we practice the possible into the real. It may be that the kiss in my dream was not nocturnal fantasy but, like the heron's dreams of flying, the practice of possibility. It may be that we evolved to dream ourselves into reality—a laboratory of consciousness that began in the bird brain.

Dreamwind
By Daniel Godínez Nivón

Long before the first breath of life stirred, the Earth was already dreaming. The sky had not yet opened, the oceans lay unborn. Still, the planet pulsed—rock, water, lightning, and magnetism weaving the conditions for what was to come. It was a kind of primordial dreaming, rehearsing the rhythms that would shape all things. If the world itself dreams, what do birds inherit? What songs rise from this ancient sleep?

1

Dust and rock drifted in perfect silence. Fragments of ancient stars collapsed under their own weight, drawn together by gravity's pull. The heat was merciless, churning stone into magma, forging a young Earth.

As its body grew, the movement of fragments imparted a slow and uneven spin, until the planet found its cadence—a fragile balance between attraction and motion. This primordial pulse, from which everything is born, gave rise to the first cycles of light and shadow, dividing time into day and night. Each rotation of the Earth was a gesture of becoming, the first breaths of a living, dreaming world.[1]

In these first breaths of the atmosphere, the oldest clouds combined with heat and gases rising from the Earth's depths, generating surges of energy that gave rise to the first lightning. For millions of years, these flashes lit up a turbulent sky and unleashed forces that repeatedly reshaped the Earth's surface.

2

Inside, minerals rearranged themselves under the invisible forces of heat, time, and the stresses of the Earth's crust. This inner world of dynamism held ancient dreams in its crystals and crevices. Each crack was a geological memory, an echo-in-becoming of this planet taking shape.

When water appeared, it was an agent of change. Flowing through fissures and basins, it transformed landscapes, eroded rocks, and transported minerals. Through its cycle of evaporation and rain, water wove connections and brought regeneration, supporting the conditions for future life.

Molten iron swirled deep within the planet's metallic heart, generating a magnetic field that enveloped the Earth. This invisible force shielded the forming skies, guiding the rhythm of rain and the breath of clouds. A hidden pulse, a silent thread weaving together core and sky—a rehearsal for collective life.

1 Sidarta Ribeiro describes how the cycles of light and darkness, resulting from the planet's rotation, were imprinted onto the biochemical cycles of unicellular organisms around 1.8 billion years ago. Thus, the Earth's dreams began to intertwine with those of life.

In an endless cycle of transformation, the Earth prepared for collective dreams long before it was born.

3

When I was a child, I used to believe that swallows lived within rocks and beneath the earth. I imagined them inhabiting the mountains, shaping their very structure. I vividly recall an evening in Ocosingo, Chiapas, when I witnessed a murmuration of swallows for the first time. It was a moment of indescribable awe: the sky darkened, and the birds, moving as one, transformed into a shifting shape of wind. As they each disappeared into the whole, my eyes couldn't rest on any single one of them, as though my gaze were being carried along a river, full with swollen droplets after torrential rain.

My family and I, along with other neighbors, stood at the entrance to an underground cave, which was more like a cliff exhaling life into the mountain. In a single breath, all the swallows in the world plunged downward, a dark waterfall folding into the earth. The night itself seemed to have taken flight, only to return home. This moment left a profound mark on me. Ever since, whenever I see a mountain, I imagine its insides filled with birds.

Birds and mountains seem to share an intrinsic bond, one that science is only beginning to unravel. There is a theory that migratory birds possess an exceptional ability to navigate the planet due to small amounts of copper in their heads, magnetite in their beaks, and cryptochromes[2] in their eyes, which allow them to perceive the Earth's magnetic field as visual patterns. Perhaps, after all, they carry a piece of the mountain within them, bearing minerals and memories that resonate with the Earth's strata.

In this way, the relationship between a murmuration and a mountain can be seen as the manifestation of a deep dream—a dream of the Earth that grows and expands in complex cycles. Each mountain is a prolonged thought; each bird a memory connecting the present with the oldest layers of the soil. In and on a mountain, everything is connected: the light filtering through the canopy, the leaves falling to nourish the ground, the fungi communicating between distant trees through their subterranean networks. Everything participates in a shared dream, a continuous rehearsal of life.

4

Birds were singing long before humans learned to dream, even before mammals—which at that time were quite small—roamed the Earth. Sixty-five million years ago, as dinosaurs began to disappear, birds were already dreaming. What songs resonated in those dreams? What echoes of the world did they share? Today, on a planet transformed by wounds and destruction, the birds are still singing. Their melodies are an invitation to listen, to imagine possible futures.

* * *

In my artistic practice, I draw inspiration from the idea that dreaming is both an act of resistance and a way to

2 Cryptochromes are light-sensitive proteins found in plants, animals, and some bacteria. In animals, they play a key role in circadian rhythms—the biological clocks regulating daily cycles of behavior and physiology. In migratory birds, cryptochromes are also associated with magnetoreception, enabling them to perceive Earth's magnetic field as visual patterns to aid in navigation.

foster relationships across time. From an Indigenous perspective, dreams are portals to collective knowledge, offering a space to imagine sustainable ways of inhabiting the world. Dreams are not confined to individual experience but connect human and non-human worlds in a dynamic interplay of transformation and agency. Colonization, however, disrupted these worldviews, severing deep ties between dreams and the land, and diminishing all recognition of the shared agency that sustains social life. My work seeks to counteract this disconnection by reaffirming the ethics of *tequio*,[3] emphasizing mutual care and responsibility.

A pivotal influence on my practice has been my collaboration with traditional Triqui midwives from San Juan Copala, who, like birds, receive knowledge through dreams. Their teachings reveal how certain skills and knowledge are not learned in waking life but are gifted in the oneiric world. These midwives have shown me the importance of sustaining an ongoing dialogue with dreamscapes to access and preserve collective knowledge. Together, we have developed long-term collective dreaming processes, which we call Oneiric Assemblies, to explore the plant forms that emerge in dreams and to rediscover ways of dreaming collectively. These experiences highlight the vital role of dreaming as a communal

3 *Tequio* is a traditional Indigenous practice central to many communities in Mexico, particularly among the Zapotecs, Mixtecs, Mixes, Triquis, and other groups. It refers to a communal work system based on reciprocity and collective responsibility, where members contribute their labor for the benefit of the community, through work such as building infrastructure, maintaining public spaces, or supporting communal projects. *Tequio* embodies an ethics of mutual care and shared agency, reinforcing social bonds and ensuring the sustainability of communal life.

practice, connecting us to shared sources of strength and knowledge while fostering pathways for mutual care and resilience.

During my residency at the Jan van Eyck Academie (2021–2022) in the Netherlands, amid the coronavirus pandemic, I spent time observing birds from my studio. I studied their songs, their relationships with their environment, and their migratory patterns. Their connection to dreams and singing became, for me, a bridge to San Juan Copala and the traditional midwives of Mexico. Both birds and midwives learn from dreams, and both dreaming and the practice of midwifery serve as sources of knowledge that help hold the world together, nurturing and caring for it and the life it holds.

Researchers at the University of Buenos Aires, led by Dr. Gabriel Mindlin, have discovered that birds dream and, during sleep, "sing in silence." By monitoring the muscular contractions of their vocal tracts, they found that birds rehearse their songs in dreams. This insight expands our understanding of birds' dream-practices, revealing how these silent songs weave the individual into the collective. In a world reshaped by environmental crises, exploring the processes of such adaptable beings may offer clues to resilience and survival.

Dreamwind is the result of this study: a piece that seeks to conjure the oneiric songs of migratory birds and their persistence in the face of environmental crisis. I like to imagine how birds' songs weave the world together. I think of them as seekers of new melodies to care for the world, resist its destruction, and create paths toward possible, liveable futures. Their songs are acts of healing and balance, tasks to be shared among all living beings.

If we listen—truly listen—to birds and the ancient dreams that shape the world, we may still recover the songs needed to weave a world not only habitable but full of wonder. Like birds, we must learn to sing—even through our nightmares—so the world may be born anew, again and again, carried on the breath of our dreaming. Their songs invite us to dream together, to seek melodies that heal and nourish, that call new life into being. In this way, birds can help us remember how to dream, and how to live, together.

THE SPIRIT

Sara Sejin Chang (Sara van der Heide)

This painting is part of a series created shortly after Sara Seijn Chang (Sara van der Heide)'s graduation from De Ateliers in 2001, in which the worlds of the living and the deceased are portrayed as deeply intertwined. *The Spirit* was inspired by a newspaper photograph of a man with a beard, glasses, and a chicken on his head. In the painting, a towering figure holds another being, while a smaller figure is seen at the level of the larger figure's stomach, gazing at its reflection in the water. These forms convey a transcendent connection, suggesting existence beyond the physical. The piece is constructed with thin, translucent layers of oil paint, a technique that enhances the interplay between presence and absence, surface and depth, existence and nonexistence, reinforcing its spiritual and ethereal essence.

Sejin Chang (van der Heide) creates works that blend spiritual evocation, historical research, and the deconstruction of colonial narratives. Her art functions as an act of historical repair to find healing and belonging. She challenges Eurocentric systems of categorization and racialization to reveal their pervasive influence on life and contemporary Western society. Her work is both poetic and intimate, proposing a meta-cosmic, inclusive approach to modernity—one that redefines notions of value and time while transcending personal and biographical boundaries.

Chapter 2:

Identity and Imagination

(49)

(50)

(51)

(55)

(56)

(57)

(58)

(64)

(62)

(63)

(69)

(70)

(53)

(54)

(60)

(61)

(67)

(68)

(73)

(72)

(74)

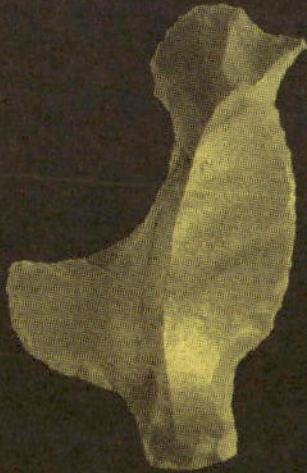

(75)

(49)

(55)

(62)

OISEAUX DU CONGO
4. — L'Aigrette Garde-bœuf et le Martin-pêcheur
HARICOTS A LA TOMATE LIEBIG: plat préparé complet
Reproduction interdite
Explication au verso

OISEAUX DU CONGO
5. — La Grue couronnée et le Serpentaire
DOUBLE CONCENTRÉ DE TOMATES LIEBIG: Purée 100 % italienne
Reproduction interdite
Explication au verso

OISEAUX DU CONGO
6. — Le Tisserin et le Nectarin
EXTRAIT DE VIANDE LIEBIG: pour la bonne cuisine
Reproduction interdite
Explication au verso

(70)

(69)

(73)

(72)

(74)

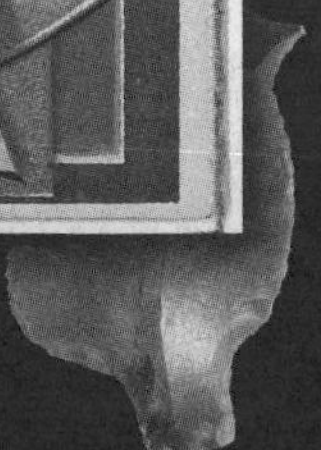

(75)

(49)

(55

(6

(69)

(53)

(54)

(67)

(68)

(72)

(73)

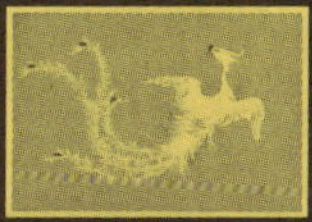

(74)

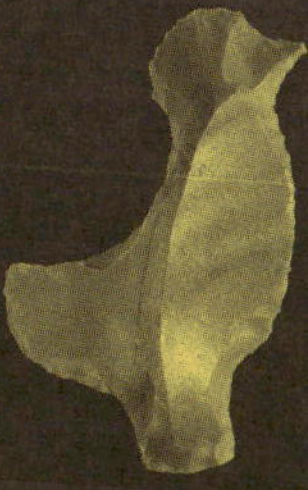

(75)

(49

(55

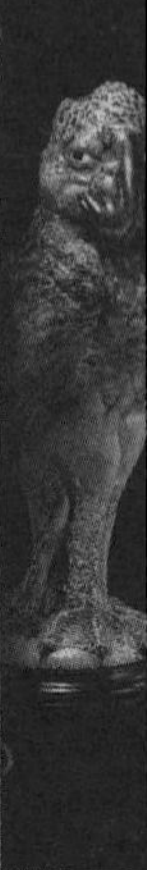

(69)

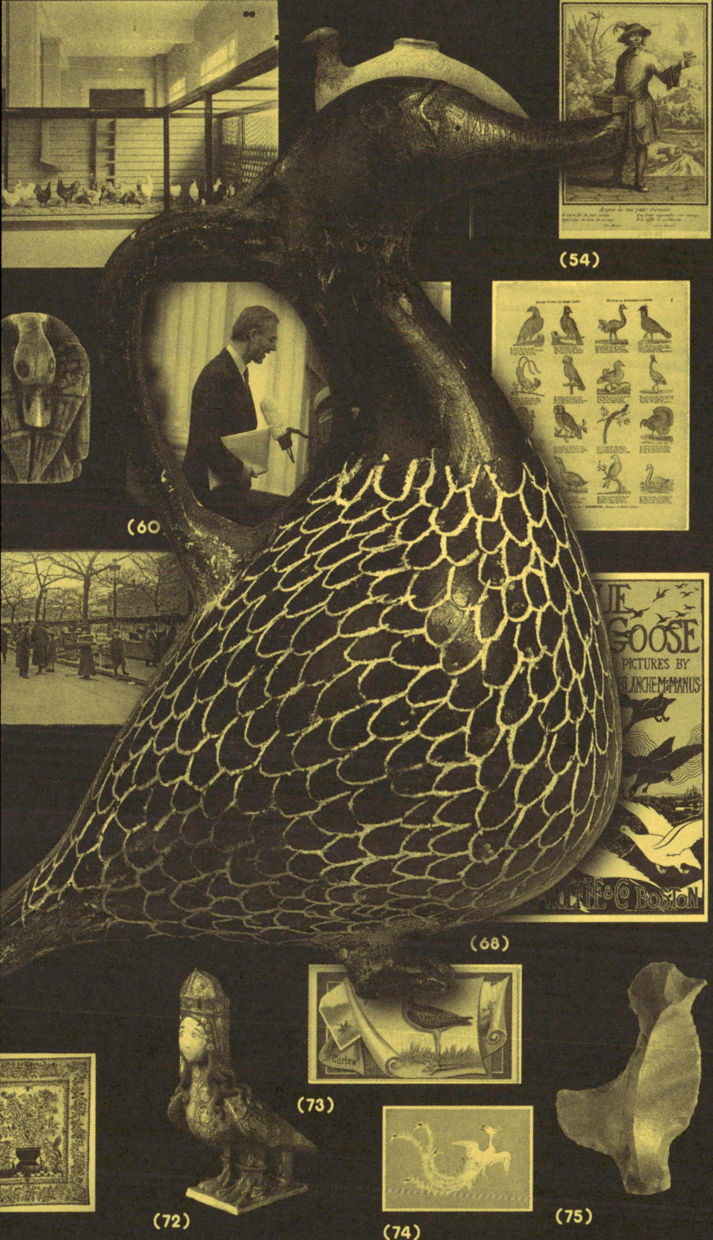
(54)
(60
GOOSE
PICTURES BY
BLANCHE McMANUS
BOSTON
(68)
(72)
(73)
(74)
(75)

(49) (51) (55) (58) (62) (69)

(60)

Argent de mes petits Oiseaux
Il est si sot de son oiseau
Qu'il vient de tirer de sa cage
Que pour apprendre son ramage
Il le siffle en godelureau.
Chez J Bonnart
au Coq avec pri/.
G157128
06223

(72)

(74)

(75)

(49)

(55)

(62)

(70)

(68)

(74)

(75)

HOW TO COCKATOO
BY SERGIO ROJAS CHAVES

How are national symbols formed, and what happens to a country's image when its national symbols change? For years, Costa Rica was "symbolized" by two animals: its national bird, the Yigüirro, and its national animal, the white-tailed deer. But over the last ten years, that number has grown from two to six. Are there still more to come? What are the motivations for creating these symbols, and how do they affect our national identity?

Let's imagine a non-native species were to be introduced into the country's imaginary by the tourism industry. This species would appear frequently in the public realm; we would first begin to see it on souvenirs, crafts, and advertisements, and then, over time, even on government literature and banknotes. Might we come to believe that this species had always been a part of our environment? Would we remember it being present where it never really was?

Artist Sergio Rojas Chaves has been approaching these questions with a particular focus on cockatoos and their presence in Costa Rican souvenir shops. In response to mass tourism, the Costa Rican souvenir industry has shifted from selling locally handmade crafts to re-selling objects imported from China. Souvenir shop owners are able to order customized products inscribed with the words "Costa Rica" from many large-scale manufacturers via online platforms for much lower prices compared to the price of local production. As a result, Costa Rican artisans have largely been pushed to the side; many have abandoned their craft, while younger generations are not interested in continuing local traditions as they see limited profits and opportunities to sell.

Amongst the options on offer from these souvenir factories are "tropical bird packages" that include figurines, magnets, and beach towels depicting different kinds of parrots, peacocks, flamingos, and cockatoos. In order to maximize their profit, shop owners will sell souvenirs depicting all of the bird breeds that come in their pre-ordered packages, regardless of the existence and relevance of the species in the country of resale. These mass-produced souvenirs also mimic traditional craft styles from other countries in the region like Mexico, Guatemala, and Perú, which creates a bizarre amalgamation of representations and handicraft styles that, to the untrained eye, may appear to be locally produced.

Cockatoos are actually native to Australia and Oceania and, like parrots and macaws, have been popular as pets—a practice rooted in colonial extractivism—since the 19th century, due to their striking plumage and sociability. It is common to find images of cockatoos on tropical-themed European and North American decorative products such as fabrics, wallpapers, and porcelain figures, however, functioning purely as an image in these contexts, these birds are divorced from their natural habitat, leading to confusion about their origins.

During a residency period at salita_temporal in San José, Costa Rica, Rojas Chaves worked on a series of sculptures resembling three-dimensional collages, in which he transformed everyday objects into cockatoos. While cockatoos are not native to Costa Rica, the works offer an invitation to imagine a not-so-unlikely future where the image of the cockatoo is omnipresent in the country. He presents an interpretation of cockatoos "handmade in Costa Rica," a nod to both the souvenir industry and the possibility of non-native national symbols. This series of works provides a space to question the neo-colonial aspects of tourism and the complicity of Costa Rican society in perpetuating the exotic gaze.

Rojas Chaves sees this gaze as a perspective coming from outside, crossing geographical or cultural boundaries, and taking hold. The act of exoticizing depends on the maintenance of boundaries that preserve cultural difference and distance to evoke a sense of astonishment and wonder in the beholder. Like a cage that is looking for a bird, this exotic gaze renders people, objects, and places strange even as it domesticates them, effectively manufacturing otherness while fetishizing a supposedly inherent mystery.

Montage of the sculptures made during the residency, featuring *ChanclaCacatúa, SillaCacatúa, BolsaCacatúa, BancoCacatúa, CarteraCacatúa, PlatoCacatúa, MacetaCacatúa, ChoneteCacatúa, CanastaCacatúa, and CornetaCacatúa.*

① Poster image made as a reaction to seeing cockatoo souvenirs in Costa Rica for the first time. Translated from Spanish: "there are no cockatoos in Costa Rica."

② Small beaded cockatoo found in the central market in San José.

③ Poster generated using Open AI's DALL-E 3 image generator, using the prompt "tropical tourism poster of Costa Rica featuring a cockatoo on the beach."

④ Photomontage of cockatoo souvenirs found around Costa Rica.

⑤ Poster generated using Open AI's DALL-E 3 image generator, using the prompt "tropical tourism poster of Costa Rica featuring a cockatoo on a banana tree."

⑥ Cockatoo on an official tourism advertisement from 1974, designed by Swiss artist Jean Pierre Guillermet for the Costa Rican Tourism Institute (I.C.T).

⑦ Photomontage of the Summer Festival 2024 in San José. The main carriage of the parade was decorated with cockatoos, and after the parade people could take photographs with them.

⑧ Poster generated using Open AI's DALL-E 3 image generator, using the prompt "tropical tourism poster of Costa Rica featuring a cockatoo."

No hay cacatúas en
Costa Rica

COSTA RICA

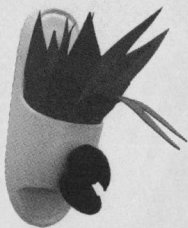

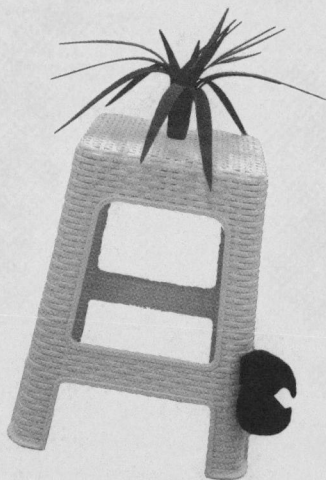

6

THE GREENER GREEN
BY MARJOLEIN VAN DER LOO

Espoo, (FI) June 7, 2020

Dear Jorge,

Someone new has entered my life and turned everything upside down. Since we last spoke, I have been "collecting" birds on the campus and by the sea. So far, I have heard or seen around thirty different species, each of which I've carefully recorded in an app. But one stands apart from the rest.

On May 9, at 1:34 AM, I woke up to a call—loud, unreal, almost dreamlike. It echoed through the night with a depth and volume I had only ever encountered in instruments or high-caliber sound installations. Yet this sound was unfiltered, unmediated—pure vibration moving through the air, weaving between tall trees and brick apartment buildings. I believe it was a male of the species; they are usually the ones who sing. His call was unlike any voice I had known, human or bird. It was both song and call, scream and growl, piercing through the long Finnish twilight, adding an unfamiliar mystery to a night that, before midsummer, never fully darkens.

The next night, I listened again and thought of how I might describe him to you. This is what I came up with: Imagine a creature who spent the '90s in a living room, surrounded by kids playing *Street Fighter*, before migrating to the jungle. He seems to embody both the edges and the heart of a rainforest. His song is a chaotic, mesmerizing sequence of gargles, rolls, roars, clicks, screams, chirps, drips, blinks, peeps, whistles, lures, and drum beats, never appearing in the same order, as though an unpredictable composition is constantly being rehearsed. The crisp, staccato quality of his sounds reminds me of retro video games, but then suddenly there will be an emotional shift—an infusion of reality that only a living, breathing musician could produce after years of devotion to their instrument. Here, which is neither a '90s room nor the jungle, but the south of Finland, they call him Satakieli—the thrush nightingale or Nordic nightingale.

I hesitate to speak of "the jungle" with you. I have never been to one, nor do I know much about them. It would be an assumption to expect that you, born and raised in Brazil with a deep interest in ecology, would have first-hand experience of the Amazon. Still, this bird, this voice, has set me on a path to exploring the idea of the jungle from my Dutch perspective, one largely shaped by the stories and artworks resulting from colonial imagination.

I'm thinking first of Henri Rousseau's paintings. They capture the sensation this bird evokes for me (*see image 1*). I was first introduced to Rousseau's work when I was around ten years old, during weekly painting classes. My teacher, Eveline, showed us an image of one of his jungle paintings, and at the time I was underwhelmed. His style seemed neither as technical as Rembrandt's nor as bold as the impressionists or pointillists I admired. Yet, even then, I felt the mystery in his vision of the jungle. Now, looking at these paintings again—even on a screen—I can sense the dampness of the soil and the thick, humid air, dense with countless scents. I can feel the soft, hairy petals and sturdy branches, the sharp leaves leaving traces on my skin, and the concave chalices holding a mixture of rainwater and tiny insect corpses, preserved for drier days ahead. This jungle hums with sounds and movements beyond my comprehension: I can only imagine.

(1) Henri Rousseau, *The Equatorial Jungle*, 1909, via Wikimedia Commons. Public domain.

But where does this vivid sensory information come from? And why am I imagining "the" jungle as singular? My imagined jungle, whose fullness saturates my mind with its restless life and lurking dangers, is shaped by secondhand knowledge; like Rousseau, I have never set foot in one. My understanding has been mediated through abstracted sources—zoos, botanical gardens, *National Geographic*, and children's books. Even further abstracted are the scientific classifications of vegetal and animal life I associate with "the jungle." These names that I once took as objective carry long histories of power and coercion. The way we name and categorize nature is deeply entangled with colonial histories of extraction and control.

During the Age of Enlightenment, the philosopher Jean-Jacques Rousseau, deeply invested in botany, described "nature" as something to be found in the colonies.[1] Centuries later, the mid-20th-century thinker Gilles Deleuze would locate this idea within the colonial imaginary, understanding European perceptions of nature as assumptions formed from images and descriptions of the Global South and colonized lands.[2] Much of my education has been built on the colonial framework—a system that studies, categorizes, extracts, and imports exotic life forms that once belonged to different ecosystems.

Recognizing the colonial gaze embedded in Rousseau's paintings makes it difficult for me to approach them with pure imagination. As art critic Roberta Smith observed, these works were emblematic of how the byproducts of French colonialism were transformed into widely consumed and provocative entertainment.[3] Simply brushing aside this layer of history would render any solely imaginative response naive. Yet, understanding this material context, I am still searching for something else in the image of the jungle—something like a way of acknowl-

1 Jean-Jacques Rousseau, *Discourse on the Origin of Inequality* (Marc-Michel Rey, 1755).

2 Ovidiu Tichindeleanu, Decolonial Summer School Lecture (Middelburg, NL, July 5, 2020).

3 Roberta Smith, "Henri Rousseau: In Imaginary Jungles, a Terrible Beauty Lurks," July 14, 2006, *New York Times*, https://www.nytimes.com/2006/07/14/arts/design/14rous.html.

edging the oppressive structures of our time while envisioning justly abundant futures beyond them.

The sound of the nightingale (which I hear here in Finland) and the thoughts of the jungle it evokes also transport me to Apichatpong Weerasethakul's film *Uncle Boonmee Who Can Recall His Past Lives* (2010), which delves into the theme of reincarnation. Though parts of the film use a constructed "cinema jungle," I experience its setting—the moist deciduous forests of northeast Thailand—as more genuine and less exoticized than Rousseau's canvases.[4] In this world, dreams and reality move and blur, especially at the edges, where human settlements meet the forest. Many of the film's shots show how thick the air is with humidity and the ceaseless calls of insects. These sounds stretch my understanding of how an insect might express its being. I reminisce specifically on one scene of the film in which a family dinner with Uncle Boonmee takes place at night. Their house is located at the edge of a forest, and apart from the silhouettes of some trees, the surroundings are pitch black and filled with the sounds of many insects. The family is seated on the veranda of the building, and two chairs at the table remain empty. Halfway through the dinner, the arrival of the spirit of Boonmee's deceased wife startles the family, but is followed by a calm and pleasant conversation. Not much later, his long-lost son appears in a non-human form and joins the gathering. The film isn't

4 Weerasethakul notes that the film depicts the memory of the landscape he grew up with. See: Apichatpong Weerasethakul, "Spotlight | Ghost in the Machine: Apichatpong Weerasethakul's Letter to Cinema," interview by Mark Peranson and Kong Rithdee, *Cinema Scope*, June 18, 2010, https://cinema-scope.com/spotlight/spotlight-ghost-in-the-machine-apichatpong-weerasethakuls-letter-to-cinema/.

concerned with how this is possible, but what the meeting means for the terminally ill Boonmee.

Weerasethakul presents a world where living and deceased, human and non-human coexist fluidly, harmoniously, even sexually. In his museum exhibitions, too, the director dissolves rigid dichotomies—nature/culture, dead/living, human/non-human, day/night—using projections, installations, and short films to craft immersive, liminal spaces. I imagine my companion, the nightingale, inhabiting such a world. Preferring to hide in dense undergrowth, nightingales are rarely seen—only heard. Their beige and brown plumage is unremarkable, yet their voices contain multitudes. Even after listening to him nightly for over a month, I hear more than just a bird in his call. Like the beings in Weerasethakul's film—humans, non-humans, hybrids, past lives—he seems to dwell both near and far; in the thicket, just beyond my reach, but ever present in sound.

I imagine that cohabiting with the forest and its creatures—welcoming the thick dark night, living and dining in its liminal sound-filled spaces—is to exist in a blurry melting pot of species and their varying degrees of mortality. But here in Finland, where I listen to the nightingale, there are no dark, damp nights; everything is a fresh green, and his presence seems to obscure my perception. Since Satakieli's arrival, the living room has taken on a green hue, reflecting the radiant wall of life just ten meters from my window. I don't know if I simply forgot how intense green can be at the start of spring or if Finland's green is, in fact, greener than others. Perhaps it's the light bouncing off the nearby sea, amplifying the saturation. This abundance of color is tangible: during video calls, when I face the window, my head on the laptop screen glows with an eerie green tint. I look like one of those late-medieval portraits where the paint's metallic white pigments have oxidized, giving the subjects a sickly pallor. Sometimes I turn my laptop around to show the others that my ghostly hue isn't caused by nausea.

A week into working the night shift, the nightingale begins to sing during the day, completely reshaping my perception of reality. Dreaming of the jungle to the sounds of his nightly calls was one thing, but his song infiltrating daylight alters my world in unexpected ways. It reminds me of the blurred sense of time I experienced during quarantine—days folding into nights, memories becoming hazy and unmoored. Like a dream, the familiar and the strange intertwine. The Finnish pine forests and birches at the coast now carry an echo of the jungle's siren call.

This bird introduced an uncanny strangeness into what I thought I knew. His presence and my growing obsession with him have become an invitation to embrace the "strange stranger"; an entity whose behavior eludes full understanding but demands engagement, as Timothy Morton describes.[5] Morton, in *The Ecological Thought*, urges us to see ourselves as inseparable from nature, rather than standing apart from it. They argue that nature is not something distant but something within, around, and between us—an interconnected system beyond our full comprehension. This is precisely what the nightingale represents: an unsettling presence that defies categorization, reminding me that no matter how much I try to grasp his meaning, I will never entirely understand him. And yet, this incomprehensibility is precisely what

5 Timothy Morton, *The Ecological Thought* (Harvard University Press, 2010).

draws me in. His song makes the everyday uncanny, stretching my perception of time and space, much like how Morton describes our relationship with nature itself.

I recently attended a storytelling workshop led by the artist Jumana Emil Abboud and organized by Casco, an arts institution in Utrecht, NL. In the session, Abboud shared a fable about a young girl outsmarting a mighty tiger, and then reflected on the power of storytelling to shape our understanding of the world. She spoke about recognizing the heroine within our own lives, using abstraction and exaggeration to turn everyday encounters into narratives that reveal deeper truths. Perhaps this is what storytelling does—it allows us to sit with the "strange stranger," to engage with the unknown rather than try to control it. Her words resonated with me as my growing fixation on the nightingale has started to take on a shape beyond mere observation. The nightingale, like a fabled creature, eludes capture. And maybe that's the point. To listen, to imagine, to accept that some mysteries are not meant to be solved but *lived*. This bird, with its disruptive presence and elusive nature, has become part of my own unfolding story—one that dismantles the borders between dream and reality, the known and the unknowable, much like Morton's strange stranger.

During lockdown, I spent late spring wandering the bird sanctuary around the campus where I lived, growing attuned to the birds inhabiting the area. Immersed in the dense green, listening to their songs, I experienced a dreamlike estrangement. There was always a sense of anticipation—when would the nightingale join the ever-changing chorus? Unlike blackbirds, finches, or willow warblers, the nightingale seemed to operate in shifts. Like me, he didn't sleep through the night. Yet even during the day, his song could surprise me, slipping in between ordinary sounds—water running in the kitchen might suddenly transform into the trill of the gale. At night, his song merged with my dreams, an overlapping of sensory worlds where I could no longer tell where my imagined green thickets ended and his real voice began.

The thrush nightingale, the one I have been following, is distinct from but closely related to the common nightingale. They resemble one another visually, yet their sonic differences are striking. The Dutch Bird Protection website describes the thrush nightingale's song as "explosive, with large crescendos, thunderous and more powerful" than that of its relative. It prefers to stay hidden, though not with such dedication as its southern kin, and favors areas near large bodies of water. It thrives in places that turn lush and green after heavy rain. Check.

[2] Christof Bobzin, *Drawings of the Thrush Nightingale*, 2008, via Wikimedia Commons. Public domain.

But what makes it Nordic? These birds only spend about sixty days of the year breeding in North-East Europe and temperate Asia, then migrate south, stopping briefly in Ethiopia and Zambia before continuing to Southeast Africa. For the remaining eight months, they are travelers. Should they really be called "Nordic" when they spend most of their lives elsewhere, or rather "nomadic"? Or, rather than a Eurocentric classification, is the location of their breeding ground the guiding factor in their taxonomy? What names are these birds given in the languages of Southeast Africa? With extreme changes in global climate materializing dramatically in parts of Africa, many migratory birds are struggling. Warming temperatures are disrupting established ecosystems, and migration routes are shifting in response. The song I hear in the Finnish twilight might already be a fading echo of a landscape that is disappearing in Zambia. The nightingale's song here is not just seasonal but a reflection of a planet in flux.

Before I ever heard or saw a nightingale, I had heard stories about them as a child. Are you familiar with Hans Christian Andersen's *The Nightingale*? A Chinese emperor, moved to tears by a wild nightingale's song, receives a jeweled mechanical bird that sings to him nightly, replacing the real one. Eventually, the artificial bird breaks, and years later, as the emperor lies dying, the real nightingale returns. Its song is so powerful that Death withdraws, sparing the emperor—on the condition that their bond remains a secret. Having spent so much time listening to nightingales, I now understand why myths attribute such magical abilities to their songs. These stories confirm their voices as something extraordinary, weaving didactic lessons into nocturnal melodies that keep children and adults awake during bright Nordic summer nights. If birdsong can stave off death, what other ailments might it heal?

(3) Edmund Dulac for Hans Andersen, *The Nightingale*, 1911, via Wikimedia Commons. Public domain.

I heard that during lockdown, many people began to remember their dreams. Eva Meijer writes that "during sleep, one practices freedom."[6] The constraints of modern life have largely suppressed this ability. But now, a space for imagination has reopened for me. Listening to the thrush nightingale and envisioning this feathered stranger has led me to daydream and construct a story for you in these confusing times. In the storytelling work-

6 Eva Meijer, *De Nieuwe Rivier* (Das Mag, 2020).

shop, Abboud spoke about how fables allow us to transform everyday encounters into something larger than life. Maybe this is what I have been doing all along—turning my sleepless nights, my green-lit days, and this bird's ephemeral presence into a story that helps me make sense of the world.

If the nightingale in Andersen's story could hold the power to drive Death away, who's to say that the song of a bird cannot also restore something lost in us? Perhaps, like Abboud's heroine, I am learning to outsmart uncertainty, navigate the strange, and embrace it, even as it remains unknowable.

Do you know any fairy tales or myths about nightingales? Or other birds with extraordinary songs? What heroes have you encountered this spring? Remember, in fairy tales, anything is possible.

With love,
Marjolein

THE SOCIALITY OF BIRDS

Reflections on Ontological Edge Effects

ANNA LOWENHAUPT TSING

N THE ISLAND OF WAIGEO, off the coast of West Papua, Indonesia, I've been interacting with a community currently fascinated by birds. Birds, people figure, might be the path to the future: they might attract international tourists who could fund village enterprises. And yet relationships between people and birds seem to me filled with ambivalence. Only the pagans of the past, local residents assured me, had special relations with birds and other animals. Today everyone is a Christian, and the terms of Christianity, they told me, require a disavowal of connections with other beings. This seemed to me a puzzle worth attention. Given this disavowal, why do so many men notice birds, and with such precision? Why are Waigeo men such good field guides for international bird-watchers? Earlier anthropologies of bird-watching, including in Papua (West 2006), have stressed the cosmological gulfs dividing Indigenous residents and Euro-American experts. How is it, then, that they can talk to each other at all?[1]

To answer this question requires bringing the sociality of birds into the conversation. In doing so I draw on the pioneering work of Deborah Bird Rose. Rose changed how practitioners get to do anthropology. She urged ethnographers into multispecies *attention* (Rose 2011). She showed us how animals participate in the social and ethical lives of human beings and that human responsibilities and modes of empathy stretch beyond humans to include the social responses of other beings. Despite the challenge of even greater communicative gulfs, my questions cannot be limited to interactions between one kind of human and another kind of human. Varied kinds of birds respond to varied kinds of humans, and vice versa (van Dooren 2014). Indeed, including the birds might allow analysts to glimpse overlapping strategies for empathetic attunement, despite cosmological gulfs, not only between one bird and another but also between expert and vernacular and between Euro-American and Papuan. I do not deny the colonial heritage of European bird science (e.g., McGhie 2017), but I do not stop with its reaffirmation. In dialogue with Rose (2004), I follow call-and-response as it refuses to stop at the lip of ontological ravines.

One of the most memorable sections, for me, of Rose's book *Reports from a Wild Country* is her exploration of Australian Aboriginal relations to the cattle brought by white settlers. On the one hand, Rose is clear about the colonizing role of cattle; cattle are "four legged soldiers in the army of conquest" (2004, 86). Furthermore, the celebration of cattle in white Australia is a performance of Man's triumph over Nature. On the other hand, Aboriginal cattle handlers manage to do something surprising with cattle. "Cattle events, rather than performing triumph, actually perform uncertainty, and thus contribute to the ongoing life of the Year Zero as a manifold of possibilities [i.e., Aboriginal ways of understanding time and community, with their contrast to white progress narratives]" (89). In rodeos, cattle respond to both settler and Aboriginal cultural logics. Rather than stopping with ontological differences, Rose brings us into performance events that sponsor multiple forms of relation, human and not human. This practice inspired me to open the study of birds in Waigeo beyond the otherness of international bird-watchers, as established in anthropological literature. How might local residents and international bird-watchers each, respectively but also in dialogue, respond to birds—and birds to them?

To begin to answer this question, I enlisted the help of ornithologist Kristof Zyskowski of the Yale Peabody Museum and bird photographer Yulia Bereshpolova. They showed me birds I would otherwise never have seen or heard. Kristof's and Yulia's patience and ability to notice, in turn, allowed me a much richer conversation with local residents who also saw and heard these birds, although not always on the same terms. Between these various interlocutors, the birds started to come to life for me as interlocutors on their own. Let me begin again, then, with birds.

Red bird of paradise (*Paradisaea rubra*): At dusk, they gather at the lek, a great tree the birds have designated as the place for performance.[2] This is not the place they eat, nor a place to make nests; this is the place for dancing. The males are superb in their maroon mantles, as saturated with color as velvet, with short golden capes and emerald face masks. This evening no females show, and the males entertain each other, working out their dance steps. First they shiver, shaking their feathers and then opening their wings, which turn crimson and translucent in the slanting light. Then they "moonwalk," shifting forward but pushing back in a shuffle. If the day is overcast, local people tell us, they don't perform. Yes, says our ornithologist, a study has been done showing that the birds recognize those bright days in which their colors glow. Now three males have lined up along a diagonal branch. First they shiver and shimmy; then they moonwalk. They approach the synchrony of a Motown trio. For me, this is the lasting image, those three, moving together through their steps. In the morning, there are females, and the action is less coordination among males and more competition for females' attention. And yet—we pay too much attention to functional goals. The three males dancing in the evening light still haunts me as an image of masculine beauty and coordination.

Across much of Papua and New Guinea, men adorn themselves with feathers to take on the beauty of birds. Post-Enlightenment Europeans, in thrall to the imagined force of masculine rationality, forgot about male beauty; it takes a visit with Renaissance paintings of saints to remember the celebration of male bodies in Europe. But in New Guinea, masculine beauty is paramount. When I saw red birds of paradise, I thought I sensed why: the birds astonish and amaze. In some New Guinea villages, male dancers wear birds of paradise on their heads (Kirk and Strathern 1993). In some, dancers evoke birds, at the edge of becoming them (Schieffelin 2005). In myths, birds become people, and people become birds (West 2006). Alas, none of this is true on the island of Waigeo, the home of red birds of paradise and the place I had gone to study people and birds. In the villages I visited, Christianity had plowed a great furrow separating the people who could change into birds and animals from the people of faith. The former are the "lost people," pagans now transformed into invisible cannibal witches. Christian conversion did not wipe out the pagan past, but it transformed what it means to be human so that animal transformations are signs only of malevolent, satanic powers. The beauty of birds exists across an unreachable divide. And yet: even in this echoing rift, there is an opening. Local men have not forgotten birds, since they encounter them every day in the forest. Most men know dozens of birds' sounds, characteristics, and habits. When ornithologists and bird-watchers showed up in the area to make their bird lists, locals were intrigued—and ready to become facilitators and guides.

This chapter explores the strange spaces of partial contact among birds, Waigeo islanders, and traveling bird-watchers. To understand these spaces, I've borrowed the ecological term *edge effect*, which refers to what happens in the boundary zone between habitats. My habitats here are world-making projects, the distinctive features of which some anthropologists call *ontological* (Holbraad and Pedersen 2017). At the edges of these world-making projects, contact can create unexpected effects. While we may not be able to transform ourselves into radically different others, we can learn about them from immersion in these edge effects. I'm interested in contacts among different kinds of human projects—as when missionaries and villagers together manage to call up cannibal witches. I'm also interested in the contacts between human world making and bird world making. To emphasize that element, I've divided what follows into three parts: birds watching humans, birds and humans watching each other, and humans watching birds. Designs for living together should take all three seriously.

BIRDS WATCHING HUMANS

Watching is a shorthand here by which I really mean "attending to works and lives." Birds pay lots of attention to human projects, although not necessarily in the ways we might first imagine. Perhaps the really important thing, for birds, is human infrastructure, the landscape-modification projects through which we make our living space, with major implications for all life. Many human infrastructures have been harmful for birds, as when pesticide-laced agriculture destroys their ability to hatch eggs (Carson 1962). Industrial infrastructures have been especially deadly, in part because investors do not stick around to see the results of their landscape interventions. Such infrastructures destroy not just individual birds but the very possibility of life. I think it's important, however, to distinguish industrial projects from all human existence. Thus, I begin more gently with the possibilities of humans, from birds' perspectives, by situating birds in the small villages and gardens beside Mayalibit Bay, Waigeo.

Radjah shelducks (*Radjah radjah*) are big black-and-white ducks often seen waddling around in pairs or small groups; they forage for mollusks, algae, insects, and sedges. I started this project because, on a previous research trip to Waigeo, I was surprised to see Radjah shelducks walking confidently about a village, vocalizing loudly to each other—and then again in another village. Why weren't people eating them, I wondered—or at least scaring them away? On this trip, they were in every village we visited, sometimes foraging together with village chickens and dogs. To answer my original question, however, I had to learn something about the meaning of *village*. It turns out that most of the villages of Mayalibit Bay are very recent enactments of government and missionary dreams. These are not spaces for pursuing livelihoods but rather for being civilized, worshipping God, and receiving government development aid. Not eating the local ducks is part of a new way of being human, which does not depend much on local resources. This is a story worth telling for understanding the relation between local people and birds. But, first, two more things about ducks.

First, village children enjoy feeding Radjah shelducks. Children sometimes offer the ducks the leftover rice from meals. The birds learn to haunt human living places.

Second, Radjah shelducks cannot take commercial logging. The watershed around our village was logged by an Ambonese company between 2002 and 2008. A huge swath around the village is still essentially deforested, although an aggressive species of weedy vine covers everything in green. Village people explained that the ducks were gone until the past few years; without trees, they abandoned the village. This is probably because the ducks nest in trees near where they forage. Kill the trees, lose the ducks.

All this matters for thinking about the concept of village. Until the 1950s, one elder explained, this village hardly existed on the bay shore, where it is now. Most people lived inland, in groups attuned to their sago orchards and vegetable gardens. Sago was the staple, and it was eaten along with garden foods such as taro, sweet potatoes, sugarcane, bananas, papayas, and green vegetables, as well as hunted meat, avian and otherwise. Freshwater fish and shrimp were an important part of the diet, but foods from the bay were less convenient. Historically, the bay was a dangerous place, full of slaving and fighting vessels. He mimed how people approached the shore—looking right and left and creeping carefully—on those occasions they dared to search for the bay's resources. He was born in 1968 in the shore village, but his father was born in the interior. As a child, he recalls, he could not ignore a great pile of human bones, left from the wars, near the shore. Only with the coming of

Christianity, he says, were the people pulled out of the forest to the bay-shore village sites.

But even this did not last long. In the 1970s, as part of New Order Indonesia's development plan for rural areas, villages were asked to consolidate to form supervillages where the government could control the populace and locate schools. People chafed at the consolidation, which left them far from their sago orchards and vegetable gardens, and in 2001 they mobilized to move back to their previous shore-edge places. The New Order had ended, and the new government was open to decentralization. By then, their old coastal village had become a forest, the elder recalled. Following coastal fashion in this region, they built houses right into the water, on high stilts so the tide ran underneath. But this was not the current government model of civilization, and by 2004 they had conceded the necessity to move away from the shore and to build flat on the ground with cement platforms and metal roofs. This planned-looking community is the development style to which "villages" are asked to aspire. (Meanwhile, old ironwood posts, set out in the tidal mud in front of the village, have become nesting sites for singing starlings [*Aplonis cantoroides*], who line the cavities made for crossbars with fresh insecticidal leaves to protect their eggs. The nesting starlings are the first sight as a visitor enters the village from the bay.) In 2008 the logging company built a church as payment for the watershed's forest. In 2015 a cement seawall was built to keep back the tide and transform the village's shoreline from mud into dry ground. Now this is a model village: neat, modern, and well ordered. It was one of two areas chosen by Flora and Fauna International, for example, to receive aid in developing ecotourism. It is hardly a scene of "traditional culture."

Some people, the elder recalled, did not convert and move to the coast. They became the orang Gi, "Gi people," the "lost people." They became invisible (gaib); they became witches (swanggi). Gi can turn into birds and animals, but they are also cannibals and sorcerers. That kind of power is not available to modern people on the coast, who are churchgoing men and women of faith. Even though the church front is painted with enormous men with wings (the Javanese finisher copied these angels from a pattern book, people said), there is no tie between humans and birds, even in stories, they told me. The only exception my interlocutor could think of was the possibility of reading the flights of brown-headed crows (*Corvus fuscicapillus*). If a crow flies from the west and gives a special call, it means someone in the family has died or, alternatively, that someone unexpected and unwanted will come. Of the animal relations once possible, there has been a bifurcation into prophecy—legitimate and respected—and satanic force. The latter is identified with the lost people, who retain the animal relations of the region's pre-Christian heritage. Today's village dancing and drumming (with no feathers involved) were learned from Ambonese Christian missionaries and schoolteachers.

Villagers are subsidized by a variety of government handouts, including rice supplies from the "Rice for the Poor" program. Rice handouts supply most families with a staple food for two to three months, after which cash incomes become necessary to keep the family eating rice, which is imported from western Indonesia. While older people still like sago, the elder explained, the children cry for rice. Thus, everyone scrambles to pick up a little cash here and there, from working on construction projects or selling betel in the regency capital to, in everyone's dreams, opening a homestay for bird-watchers. While the village does not yet have successful facilities, it's on everyone's minds.

Villages, then, are government and missionary infrastructures—with good results for a few birds, such as the starlings who lay their eggs in abandoned posts and the ducks that wander around the village collecting scraps. Those birds who do not thrive in the midst of such infrastructures are missing; they do not participate in this contact zone. They should not be forgotten. Still, it's worth paying homage to those who have found some use for human things. For them, new affordances are gained as infrastructure is put to new uses. In villages at the edge of mangrove swamps, government-subsidized piers form landing places for migrating sandpipers, stints, and sand plovers, who rest on the wooden surface between food-gathering expeditions in the mud. Without the pier, they might not visit this place. Human things are not useful just for humans.

In gardens, the shared use of human landscape modification is even more evident. As Rose's approach allows us to see, these are storied-places for birds as well as humans, emerging out of histories of interaction and meaning making. They are places of significance and attachment, as they are for humans, even if each of us comes to know and inhabit them differently. As Thom van Dooren and Rose (2012, 2) explain, "places are co-constituted in processes of overlapping and entangled 'storying' in which different participants may have very different ideas about where we have come from and where we are going. What would it mean, in a multispecies context, to negotiate 'across and among difference the implacable spatial fact of shared turf' [Massey 2004, 3] . . . ? What would it mean to really share a place?" (van Dooren and Rose, 2).

Most gardens here are inland, several hours' hike from the bayside villages, and because of the history I've just described, they tend to be visited too rarely for careful tending. The gardens I knew were overgrown and weedy, blending in and out of the forest. What a haven for birds! The edge of the gardens is a noisy place, full of bird vocalization. In contrast, walking through the surrounding forest was often quiet; birds were less concentrated there. Birds cluster in the trees at the edges of gardens, making use of both forest and garden habitats. Brahminy kites (*Haliastur indus*) find hunting viewpoints on high trees overlooking the bright gap. Glossy-mantled manucodes (*Manucodia ater*) display their dipping flights over the open space. These birds don't care much about the garden as a garden; it's just a gap in the forest. But other birds find the garden a good place to eat, and this might introduce the conflicting perspectives—humans versus birds—that I am calling "looking both ways."

LOOKING BOTH WAYS

Eclectus parrots (*Eclectus roratus*) are the most sexually dimorphic of all parrots; eclectus males are a brilliant green, while the females are scarlet red with blue and purple. Once, Western birders thought them to be different species. Now, however, everyone seemed to know their sexual morphs, including all my companions from Waigeo, some of whom also knew the English name *eclectus*. They were so common around gardens that my birder companions stopped caring when they flew overhead or perched nearby. Indeed, they are one of the worst garden pests in the area, along with wild boar. Flying foxes, which also haunt gardens, eat only ripe fruits; eclectus parrots eat everything. Sulphur-crested cockatoos (*Cacatua galerita*) perch on high trees around gardens, but they specialize in forest—not garden—fruits. In contrast, eclectus parrots are there to eat the food raised by gardeners, and especially bananas and papayas, green or ripe.

Our host had put a nylon net around a big bunch of ripening bananas, still on the plant, to protect it from the birds, and a bright scarlet female had caught herself in its mesh. She was screaming fiercely when we moved into the shelters by the garden, and eventually our party asked that she be released. Our host did it in our honor, and Yulia gently rinsed her plumage from the feces with which she had soiled herself during her captivity. But our host grumbled. "If it was up to me," he said, "I would sentence that bird to death." As long as the bird was there screaming, other eclectus parrots stayed away, he said, so, held fast by the nylon net, she protected his field. "I would give that bird the punishment [setrap] of standing there until death," he added. Setrap, he explained, is the punishment schoolteachers give their pupils for disobedience, handed down from Dutch colonial discipline: you might be ordered to stand on one foot or stare at the sun for an hour. The violence brings orderly, civilized behavior, and the bird, he thought, deserved it. With other garden birds, however, our host was surprisingly tolerant. Eclectus seemed to him out of bounds.

Eclectus parrots get in the way of people; other birds work hard to stay *out* of the way, and for them, too, the clash of modes of watching is particularly evident.

The Waigeo brush-turkey (*Aepypodius bruijnii*) is Waigeo's most internationally famous bird, at this moment, but it is described by villagers with the banal name of *forest chicken* (*ayam hutan*); as a result, most foreign birders conclude that villagers don't know a thing about the species. (Indonesian birders, more attuned to the excitement of international guests, call the bird *maleo Waigeo*.) The bird is famous because international confirmation is new. A bird specialist from Belgium, Iwein Mauro, reported its existence based on field observations in 2002 (Mauro 2004, 2007). He found the mounds built by males to incubate eggs in the stunted cloud forest that grows at altitudes over 620 meters, that is, on the higher mountains south and east of Mayalibit Bay. This opened a bit of a mystery for him. Historical specimens were found in the lowlands, but now, it seemed, the birds' populations were limited to a few high places. After some consideration, however, he dismissed those lowland birds, which local people knew, as "vagrants." Meanwhile, his reports spread the dogma of local ignorance; based on the local name, he figured that locals mistook the bird for an ordinary village fowl.

The villagers I spoke to had different ideas. Although everyone admitted that the bird was hard to find today except in remote high places, men said that not so long ago, in the lifetimes of their fathers, the bird had been quite common in the lowlands. They thought it had moved to high country to get out of the way of increased human activity—especially after the consolidation of human settlements along the coast in the middle of the twentieth century. Before that, it had been a regular part of hunters' diets, but that was when people were still scattered in gardens inland, or so my informants said.

One of the previous lowland specimens Mauro (2007, 108) reports involved the head and the bones left from a hunters' meal. That part of the local story finds agreement on both sides. As to the historical ecology of the species, it seems important to note that Mauro invokes a strange timeline, which appears sensible only because of the political cosmology that produced it.

Mauro imagines a past in which "nature" was untouched by human effects; meanwhile, he yokes this past to a future in which villagers are likely to destroy everything immediately. Thus, the birds' distribution in the past has nothing to do with human histories. In contrast, the future is human: Mauro's first conservation suggestion is to stop local villagers from destroying the birds. (While he admits that industrial mining and logging are key problems, he never suggests that corporations be contained.) He wants to outlaw hunting (even pig hunting—and pigs are the birds' worst egg predators); only trained scientists should have access to the areas where one might find birds (Mauro 2007, 114). Nature comes to scientists pure, and yet it arrives immediately vulnerable to native depredation. This is an exotic mythology in which scientists inherit the colonial burden of saving the land from its people.

To balance out the strangeness of this cosmology, it seems sensible to attend to the stories of local hunters, whether mythical or otherwise. For this, it doesn't matter where the core nesting areas of brush-turkeys lay historically. Let's say lowland birds are vagrants. It seems likely that they have sometimes, perhaps sporadically, been frequent visitors. It also seems likely that they have paid attention to humans in a different way than the smaller, less inviting (as food) dusky megapodes (*Megapodius freycinet*), which merely melt into the bushes when humans make an appearance, returning later. The bigger, more delicious brush-turkeys, it seems, got out of the way more firmly, finding refuge in high places. Brush-turkeys no longer visit the lowlands because there are too many people there. Ontological edge effects: brush-turkeys, too, may be capable of skilled practice on the borders between world-making projects.

Mauro's recommendations for restricting villagers' access to their highland forests resonate with common parlance among international bird-watchers, who are worried about the ways local people treat birds.[3] This reaction allows me to take some distance from bird-watchers. Surely capitalists have developed more powerful ways to kill birds, mostly involving toxicity and habitat destruction. The strangeness of bird-watchers' focus on local ignorance and destruction, then, is a reminder that bird-watching as a world-making project is as parochial and exotic as that of Waigeo islanders using colonial punishments on parrots. When every cosmology is strange, we might notice edge effects—a way into others' lifeways that does not erase the effect of observation itself.

In this section of the chapter, I've tried to show the distinctiveness of world-making projects—of birds, bird-watchers, and Waigeo island residents. This is necessary but not sufficient to understand the ontological edge effects I've promised. Let me turn to the most taken-for-granted part of bird-human relations (to Western readers): bird-watching. I hope I've already made it richer by suggesting bird-watching's interplay with many projects, human and not human. Now I'll identify some dynamics at play in the intersections across those projects.

HUMANS WATCHING BIRDS

What is bird-watching all about? One element is finding empathetic attunement with another being. This is a matter of working through edge effects, that is, tentative sites of touching. In this section I'll focus on three kinds of edge effects: copying, negotiating differences, and finding overlapping curiosities. Each takes me a little further into thinking through how varied kinds of humans and birds do and do not create common worlds together. Let me clarify: I'm not longing for the unification of all worlds. But I also refuse to assume that all these worlds are autonomous and nonoverlapping. We need to see where and how touching occurs, that is, what happens in edges.

COPYING: Birds and humans each copy the other. This is really clear among the pet sulphur-crested cockatoos (*Cacatua galerita*) that people keep in Waigeo's regency capital. Cockatoos were chained to perches in front of people's houses. One woman I met had nursed hers from the bedraggled state she bought it in to a healthy shine by feeding it a human baby food of milk and rice porridge; the bird, she said, would not eat her domestic fruit. But it could call motorcycle taxis (ojek) by crying, "Ojek," and wake the children for school by calling, "Sekolah" (school). She copied its high nasal voice, and, in turn, its voice copied her copy. Copying each other's copies: here is one edge effect.

Mimicry is not limited to the relationship of pets and owners. All the Waigeo men with whom we traveled used bird sounds to call birds. Kristof used his recorder in the same way: he would record a birdsong (or, if necessary, call it up from his library) and play it back to see if the bird would come to investigate. And the birds, too, copied. The first reaction to hearing a version of their song was often to sing it out again. One of the most impressive was the hooded butcherbird (*Cracticus cassicus*), which has a complex song and mimics the sounds of other birds around it. As we walked through a forest grove thick with butcherbirds, the songs kept morphing, drawing new elements into the melody. Like mockingbirds, butcherbirds are alert to sounds around them, copying and reweaving. Copying is one kind of attunement.

Just as biologist Scott Gilbert (2012, 336) says, "We are all lichens," to highlight our symbiotic natures, when it comes to copying and reweaving, we are all butcherbirds.

NEGOTIATING DIFFERENCES: It's easy to accept new information as long as it fits easily into the framework we already know. But it's really hard to even notice something from outside one's own world-making project.

After some years of attention to contact zones and hybridities, anthropologists have become obsessed with just how difficult it is to get outside one's own world and with how people just reinterpret what they experience—however divergent—into their own frames (Viveiros de Castro 2004). This is especially true, my colleagues say, for scientists and conservationists (e.g., Blaser 2009). I'm investigating edges because I think anthropologists have gone too far now in ignoring them, and yet the new scholarship is a helpful reminder: both disconnections and connections matter. Deborah Bird Rose motivates just this remembrance by attending to multiple ways of knowing and being while insisting on overlaps, dialogues, and cross-pollinations.[4] Paying attention to what happens at the edges of discrepant world-making projects requires noticing refusals to hear each other as well as unexpected forays into each other's worlds.

I thought about this problem a lot in my interactions with villagers, birdwatchers, and a tiny, bright bird, the yellow-capped pygmy parrot (*Micropsitta keiensis*). It was one of my most breathtaking glimpses of bird life. The tiny birds were clutching an arboreal termites' nest, near the lip of a hole near the bottom. At first I saw one, then two, then three. They glowed green, and the male had a red chest. They moved to a branch. He sat between two females exchanging overtures and glances back and forth, seemingly courting both. It felt intimate to see this, so near through my binoculars. Kristof checked his book, which said they are communal breeders who excavate in the termites' arboreal nest to lay their group's eggs.

Kristof asked Pak Noh, the village man who was with us, to climb the tree to see if there were any eggs in the hole. There were not, and besides, Pak Noh claimed, the hole was not a nest. The birds were eating there, he said. But Kristof said that parrots only eat fruit; the hole was a nest. We were at an impasse. Later, looking at another source, I read that pygmy parrots are eaters of lichen and fungi, not fruit. (Some sources said the sexes look the same, which was not true for ours; not everything one reads is true.) But might termites' nests be a good source for fungi? I don't know and can't make a judgment. But, as an anthropologist, I was interested in the conflict. Pak Noh said that parrots and kingfishers do make nests in termite constructions, but not this parrot. Kristof said Pak Noh lacked the experience to know.

Where does certainty come from? Birders come to the field expecting to identify a known series of birds. Birders do not expect to see any birds that are not cataloged in their books, and they trust the book's knowledge. Having more experience with fungi, I expect many organisms to be unclassifiable by eye at the species level as well as sometimes to lack names or classifications at all. Birds are different. Kristof and Yulia thought the lizards and butterflies they photographed might be "new to science." But not birds. Because of the international birding community, this is well-trodden territory. At least for birders, the classification of birds and the general outline of their habits are considered well known and stable. Birders add information from local people but only where it fits known gaps in their already established knowledge base.

When Kristof and Yulia, who have been around the world watching birds, explained to me that everywhere they went, local guides showed them birds, I found this a great mystery. Why would local people necessarily be interested in birds at all—and when they were, why would it be on close-enough terms to those of Western bird-watchers to allow communication? I don't know the answer for the world, but from working with villagers and bird-watchers in Mayalibit Bay, I have a few ideas about that place.

Consider the ambitious young villager I'll call Yosep. Yosep aspires to open a homestay for international bird-watchers, and he has already built a toilet and a gazebo in an auspicious place, close to the water so tourists won't need to hike. To augment his skills, Yosep—who knows no English at all—sat down with a booklet on Waigeo birds produced by a nongovernmental organization (NGO) and memorized eighty-two English bird names in one week. For me, who before this project knew almost no bird names, this seemed astounding. But without more training in English, Yosep was unable to say the names in a way that an English speaker might understand without a whole lot of work. Kristof and Yulia were annoyed by his attempts to say bird names in English. They were particularly disdainful of his attempt to pronounce *yellow-capped pygmy parrot*, which came out as a slurred single sound arranged around a two-syllable *cap-ped*. And yet—Yosep knew the sounds and habits of every single bird we heard calling while we walked together. His detailed knowledge of bird life in the forest will make him an ideal guide.

Two cultural legacies come together here. First, there is the strangeness of scientific ornithology, which is much influenced by the lay practice of bird-watching: as a result, and in contrast to the science of fungi, for example, cryptic species and traits not picked up by human eyes and ears are not particularly important in discussions of birds across the lay-scholarly line. This makes it easier to find resonance between villagers and international birders. Both international birders and Waigeo villagers know birds through sounds and appearances.

Second, there is the strangeness of Mayalibit Bay village life, which is torn between the earnest adoption of an evangelical Christianity and the still-necessary practices of a forest and bay-edge livelihood. Village men see and hear birds every day, and they continue to attend to the details of bird lives. But they are cut off from bird worlds by Christian dogma, which demonizes human-animal boundary crossing. The vehemence with which villagers denied speaking to animals was impressive; it is this refusal that makes one civilized and, indeed, now, properly human. And yet the men *know* birds. In the midst of this tension, along come international bird-watchers and a potential source of new income. The eagerness with which village men embraced this possibility is not just about the money. They have skills and experiences that keep birds important—but difficult to fit into current cosmological practice. Yosep's eagerness to learn English names is exemplary here. In the gap between Christian faith and forest practice, local language is inadequate. English names and international bird-watching practices jump in as a promising alternative. Here, both refusing and grasping the framework of others gain traction—creating unexpected edge effects. This is a difficult and risky road to navigate.

FINDING OVERLAPPING CURIOSITIES: Let me end this section with something gentler—and one more bird story, for the Raja Ampat pitohui (*Pitohui cerviniventris*). Kristof had wanted to see this endemic bird, which is special because, at least if it is like its relatives on the mainland, its feathers and skin are poisonous, discouraging predators. Because we had not encountered these birds in the villages of Mayalibit Bay, he asked Yakub, a local and self-trained bird guide who worked out of an ecotourist-oriented coastal village. Kristof told us the story of how the poison became known to Western science, itself a story of overlapping curiosities. A researcher studying Papuan birds of paradise netted a pitohui by mistake. As he tried to free the bird, it clipped him. Instinctively, he put his finger to his mouth—and his lips and tongue instantly lost feeling. He realized the bird had poisoned him. He asked the local people, who told him of course it was poisonous. Following a curiosity nourished in common space, he switched his dissertation research to study pitohui, documenting several kinds. Very few birds carry poisons.

These birds eat a poisonous beetle, he found, and are able to transfer the beetle's poisons to their skins. Incubating eggs in the nest transfers enough of the poison to protect the eggs.

Yakub knew a lot about pitohui, and before long, he caught the bird's song. We stopped on the trail (a former logging road wide enough to offer good visibility), and because Yakub did not have it on his phone, Kristof played a library recording of the male's call. He put the speaker by the side of the road, and we stepped back, hoping the birds would come to the speaker. They did. Not just one male but several males gathered to find out about this interloper. We admired their burnished breasts, took photographs, and let them go about their business. Kristof offered Yakub a copy of his song recording.

Conventional wisdom has it that male pitohui come to a call to drive invaders from their territory. We attracted at least five, and if they were all on their own territories, those territories would have had to have a rather strange shape. It does not seem so far out of bounds to offer a simpler explanation: they were curious. Just as humans remake our skills for the socialities of different kinds of humans as well as birds, birds remake their skills for the socialities of humans. If we each—birds and humans—satisfied our curiosities a little that morning, it was because our respective skills were extended into the opportunities at the edges of others' worlds.

SOME FINAL THOUGHTS

Research is always a series of edge effects across human and nonhuman world-making projects. As researchers, what we learn is what our research subjects allow us to notice through the edges we mutually create with each other. This is why it is important to keep observers in the stories we tell; otherwise, we build our prejudices into our research without realizing what we are doing. Watching pitohui together with a local self-trained professional geared toward foreign tourists and their pocketbooks on the side of a former logging road built to fund a regional government geared toward ecotourist "protection" of the very biodiversity that the logging kills is so full of contradictions I can hardly begin to unpack them. Clearly, however, the pitohui's willingness to hang out in logged-over forest and come to bird-watchers' calls shapes the possibilities of interaction.

Yet pitohui responses to bird-watchers and logging roads are still an odd-looking element of analysis coming from the humanities and social sciences.

Even social and cultural analysis that highlights the importance of other living beings continues to privilege human relations with other humans. We learn that there are varied ways people make sense of and live with other organisms. We learn that human-nonhuman relations form part of human systems of power and knowledge. We learn that other cosmologies challenge the tools of Western science. Too often, the active responses of other beings are not part of the analysis—even when the whole point is to move beyond the Enlightenment-sponsored nature-culture dichotomy. Indeed, social and cultural analysts have been wary of attention to the active practices of other organisms for fear of subsumption into hegemonic scientific logics. In contrast, I argue that allowing bird responses to human projects, as well as the other way around, into social and cultural analysis opens more avenues to consider how science and its alternatives variously shape bird-watching practices.

Making a sharp contrast between local, Indigenous, and vernacular, on the one hand, and expert, scientific, and colonial, on the other, has been an important move in questioning the hegemony of colonial science. Mary Pratt's *Imperial Eyes* (1992) pioneered scholarly attention to the ways that European natural history ignored Indigenous land rights and advanced colonial knowledge and power; a host of scholarly exposés of natural history followed. Bird-watching has been closely tied to imperial expansion; this continues in contemporary imperial conquest, as when NGOs send their bird-watchers to the Iraq marshes to support the American occupation (Guarasci, n.d.). It is important to keep a critical eye on the dangerous alliances and imperial institutions that ornithologists and bird-watchers build (Lewis 2004). At the same time, merely reiterating the contrasts between local and scientific, or vernacular and expert, is not enough. Instead, I have argued that we need attention to edge effects, where touching and overlap occur even across varied projects of world making, whether those of Waigeo residents, foreign bird-watchers, or birds themselves.

To bring nonhuman responses into social and cultural analysis also opens possibilities for collaboration between anthropologists and natural scientists in moving beyond human exceptionalism—even as we continue the work of decolonizing public knowledge. I do not believe this means giving up attention to violent histories of dispossession and exclusion, but I have devoted this essay, instead, to the spirit of possibility. In the events described here, some Waigeo villagers, Western bird-watchers, a local NGO representative, and two anthropologists, one Indonesian and one not, got a little closer to appreciating more-than-human sociality, which we will need to build better alliances together.

NOTES

1. This chapter—an early missive from a continuing project—reports on material gathered in December 2017. The fieldwork also included Salmon Weyei, at the time a local representative of Flora and Fauna International, and anthropologist Hatib Abdul Kadir of Universitas Brawijaya. Each made this research possible. I am grateful to them, my ornithology colleagues (who taught me all the science here concerning birds), and the local people who facilitated the research (and taught me everything else here concerning birds). Nils Bubandt of Aarhus University got me started in Raja Ampat and shared invaluable insights. The research was supported by the Danish National Research Foundation as part of a Niels Bohr Professorship and the Aarhus University Research on the Anthropocene project it supported (https://anthropocene.au.dk).

2. About names: No naming gets at the essence of a being; at best, it can gesture in a useful direction. In the ontological ecotones that form the subject of this chapter, naming is particularly fraught, and no choice covers enough territory. To make the chapter readable by ordinary English speakers, I privilege English common names here. I supplement those names with current Latin binomials to honor the ornithology and bird-watching community that forms part of my discussion here. In the summer of 2018, I worked with Waigeo men to put about seventy-five Ambel bird names in dialogue with international bird-watcher identifications; however, an accident of the US Postal Service seems to have deprived me of that material. Those names do not form part of this chapter as originally planned. I will have to redo that research.

3. To his credit, Mauro (2007, 113) states that human predation is not currently a major threat. Many bird-watchers, in contrast, jump to the assumption that local hunting is the first concern for any bird, even where major infrastructural disturbances seem likely to be more deadly.

4. In discussing the philosophy of Val Plumwood, Rose (2013, 103) considers how insects and other beings communicate with humans, both Indigenous and settler. She argues for attentiveness: "March flies, for example, start biting across a wide area, but the meaning of that bite varies from one locale to the next. The system opens the human sensorium, extending it through attentiveness to others: jangarla trees tell what is going on under water where humans cannot stay for long; swifts tell what is happening in the upper atmosphere; march flies tell what is happening along the banks of billabongs and rivers, whether people are there or not. For humans and others to gain knowledge from tellers, therefore, they must pay active attention."

human poetry #1, 2020

daily touch, 2020

human cooking #1, 2020

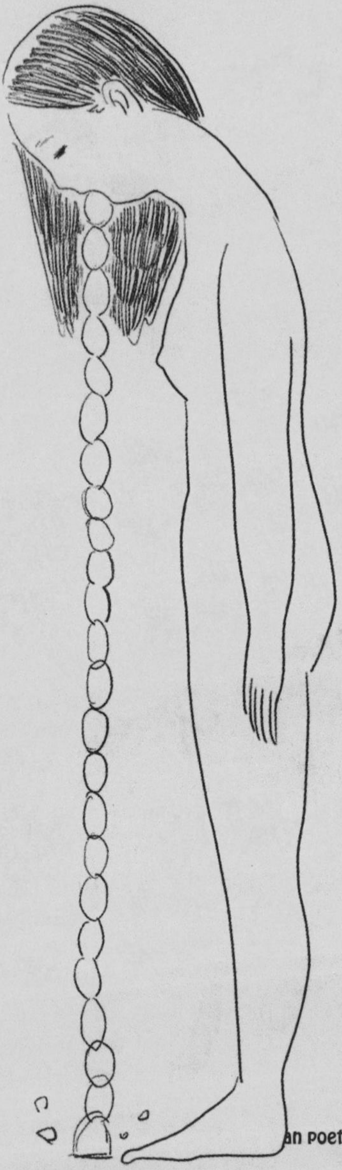

an poetry #2, 2020

gift, 2014

fountain #3, 2023

People who talking words of birds, 2020

CHALLENGING ORNITHOLOGICAL BIAS: FEMALE SONG IN ORCHARD ORIOLES

BY KEVIN E. OMLAND, KARAN J. ODOM, BERNARD LOHR, MICHELLE J. MOYER, EVANGELINE M. ROSE

Historically, elaborate advertisement traits in animals have been assumed to have evolved as a result of sexual selection acting on males (Darwin 1871, Andersson and Iwasa 1996, Tobias et al. 2011). There are, however, many examples of females displaying traits that are equally or even more elaborate than those of males across a wide variety of taxa (e.g., African starling plumage, Rubenstein and Lovette 2009; butterfly coloration, Oliver and Monteiro 2011; neotropical frog calls, Serrano and Penna 2018).

The study of bird song has been dramatically impacted by this long-standing bias toward male traits. Most early prominent studies of bird song were conducted in the temperate regions of North America and Europe, and most often by male researchers, leading to geographic and sex biases in the field (Odom et al. 2014, Haines et al. 2020). As a result, singing behavior in passerines (songbirds) has been primarily attributed to males. For example, in early editions of their widely cited book *Bird Song*, Catchpole and Slater (1995:10) defined songs as “long, complex vocalizations produced by males in the breeding season” (but see Catchpole and Slater 2008). However, this classic definition has been challenged by recent insights into song in both males and females.

A growing body of research shows that many female passerines also sing, and that many previous assumptions of absence are more likely a result of a failure to detect female song (Langmore 1998, Riebel 2003, Odom et al. 2014). In fact, phylogenetic reconstruction found strong evidence that the common ancestor of songbirds most likely possessed female song, indicating that song originally evolved in both sexes (Odom et al. 2014). Given the historical focus on male song, behavioral studies and analyses of female song are much less prevalent, especially in temperate species. In recent years, there have been numerous calls to investigate female song in temperate songbird species (Riebel 2003, Riebel et al. 2005, Odom and Benedict 2018). More studies quantitatively comparing male and female song are needed to truly understand the function and evolutionary history of this elaborate trait.

Previous studies across a variety of species have found that female song may be less elaborate in some acoustic characteristics compared to male song. For example, female Least Flycatchers (*Empidonax minimus*) sing songs with lower frequency ranges and shorter inter-note intervals than male songs (Kasumovic et al. 2003). The Montane White-crowned Sparrow (*Zonotrichia leucophrys*) female song reportedly lacks terminal elements, and sounds "softer and quavering" compared to male song (Baptista et al.1993:522). Because of such findings, female songs in temperate migratory species appears to be thought of either as rare (Morton 1996, Langmore 1998, Riebel 2003, Slater and Mann 2004) or, when present, less frequent and complex than male song (Hoelzel 1986, Baptista et al. 1993, Brunton and Li 2006, Price et al. 2009). Additionally, in temperate regions, females can be relatively quiet and visually cryptic (Price 2019), likely contributing to increased detection of and researcher bias toward male songbirds (Bennett et al. 2019).

However, a comparable number of studies have found that female song is similar to or more elaborate than male song, in both tropical and temperate regions (e.g., Arcese et al. 1988, Johnson and Kermott 1990, Pavlova et al. 2005, Price et al. 2008, Illes and Yunes-Jimenez 2009, Campbell et al. 2016, Odom et al. 2016, Reichard et al. 2018, Rose et al. 2018). In Venezuelan Troupials (*Icterus icterus*), females sing more often during the day than males, and the number and syntax of syllables is similar in female and male songs, indicating similar overall song structure (Odom et al. 2016). Eastern Bluebird (*Sialia sialis*) females sing acoustically equivalent songs to males according to five common measures of song variation (Rose et al. 2018). Recent studies have increased our knowledge of the presence of female song in temperate regions (e.g., Halkin 1997, Krieg and Getty 2016, Hathcock and Benedict 2018, Rose et al. 2018, Heaphy and Cain 2021).

In both temperate and tropical species, pre-existing scientific biases and plumage patterns may lead to misidentifying singing females as males. Female song in many tropical species can be missed as a result of males and females displaying sexually monomorphic plumage (Webb et al. 2016, Odom and Benedict 2018). In temperate species, female song may be missed because of delayed plumage maturation, when one or both sexes does not acquire their adult plumage patterns or coloration until after their first breeding period (Hawkins et al. 2012). In many species, yearling males retain immature plumage which is, in varying degrees, similar to female plumage (Lyon and Montgomerie 1986, also see Patchett et al. 2021).

Orchard Orioles (*Icterus spurius*) are a migratory, temperate breeding icterid (a member of the New World blackbird family, Icteridae) found across the eastern United States and into parts of Canada and Mexico during their breeding season (May–July). The species is sexually dichromatic, and females and older males can be easily distinguished in the field. However, males have delayed plumage maturation: yearling males have similar olive-green plumage to females, only differing from females in the black coloration on their face and throat (Scharf and Kren 2020). Most tropical icterids have female song, and ancestral state reconstruction indicates that this family historically had this behavior (Price et al. 2009, Odom et al. 2015). However, female song in temperate orioles has been significantly understudied. Male Orchard Orioles sing extensively, and their vocalizations have been described in detail (e.g., Sturge et al. 2016). There are several mentions of female song in this species (Enstrom 1992, Scharf and Kren 2020) but these accounts are brief or anecdotal. Our lab had been conducting field work on Orchard Oriole breeding behavior for over five years, but we did not notice female song in this species until our work on tropical species primed us to look for female song. These initial observations of females singing occurred primarily in the early weeks of the breeding season, suggesting that female song in this species may primarily function in mate attraction (Gahr and Güttingery 1986, Eens and Pinxten 1998, Langmore 2000, Austin et al. 2021). If this is the case, we predict that female song rates will be highest at the beginning of the breeding season, before they have found a mate, and decline after incubation begins.

In this study, we collected high-quality recordings of male and female song in Orchard Orioles, and analyzed the song structure of both sexes using eight measures of acoustic variability. Additionally, we compared song production rates from both sexes across the breeding season. To date, there have been no published detailed analyses of female Orchard Oriole vocalizations.

• METHODS •

We conducted fieldwork at 24 field sites across central Maryland (Howard, Prince George's, Carroll, and Baltimore Counties; Appendix 1). We obtained recordings for 48 males and 32 females. All recordings were made with a Marantz PMD 661 recorder and a Sennheiser ME67 or ME66 shotgun microphone with K6 powering module. We recorded orioles between 6:00 am and 12:00 pm from 28 April to 30 June 2020, and 29 April to 30 June 2021. Females were distinguished from adult males by lack of any chestnut coloring, and from yearling males by the absence of prominent black coloration on their face and throat. Sex identification was confirmed by observation of females performing sex-specific breeding behaviors, such as nest-building and incubation (Ligi and Omland 2007, Scharf and Kren 2020). We excluded songs recorded within 500 m of each other to avoid re-recording the same individual, unless multiple individuals were observed simultaneously. Given what we know about typical territory size in Orchard Orioles and our knowledge of movements of marked birds, 500 m was well beyond the distance of any likely movement (Ligi and Omland 2007, Dowling and Omland 2009).

Song rates

Once an individual was clearly visible, regardless of whether it was singing or not, we began recording for up to 10 minutes or until the location of the bird was no longer known, with an average recording length of 9 min 4 sec. All songs produced by the focal individual during the recording period (that we directly observed) were used to calculate song rates (songs/min). Multiple recordings (comprising a total time of 10 minutes or less) were combined as a single observation if taken within the same 30-minute period, beginning with the longest segment of continuous recording. Rate observations from a given location or individual were only included once within a seven-day period. One rate observation of at least five minutes from each individual was selected at random to calculate the average singing rate for each sex (male rate observations = 48, female rate observations = 27). We then performed an independent samples t-test to compare average male and average female rates (Appendix 2).

Finally, we investigated if there was any effect of time of year on the song rate of either sex. We fit linear and quadratic models to the individual-level rate data and calculated their R^2 values to determine how much variation in rate could be explained by time of year.

Acoustic measurements

We analyzed all songs in RAVEN Pro v1.6.1 (Center for Conservation Bioacoustics 2019) using a 512 pt DFT (Discrete Fourier Transform) for a frequency resolution of 124 Hz. Song elements that were at least 0.01 s apart were defined as distinct syllables (Hagemeyer et. al. 2012). We classified vocalizations as songs if they comprised at least three syllables and were at least one second apart from other vocalizations (Fig. 1). We measured nine acoustic characteristics from three songs from each individual bird (male recordings n = 30, female recordings n = 15). The three-song limit was chosen to achieve the maximum sample size of high quality female songs. Three songs was the minimum number of high quality songs that were recorded from most females. Importantly, we used the same sampling scheme for males and females, which prevents biases toward either sex.

Syllable-level measurements were averaged for each song to create a song-level dataset, and grand-means of the measurements from all three songs from each individual bird were calculated to create an individual-level dataset. The first five measurements were taken at a syllable level: syllable duration (s), minimum frequency (Hz), maximum frequency (Hz), bandwidth, and number of inflection points. The remaining four measurements were taken from the full song: song duration (s), syllable rate (syllables/s), number of syllables, and percent pause (the percentage of the song comprising silence). Minimum and maximum frequency were defined as the frequencies delineating 5% and 95% (low to high) of the spectral energy in the selected syllable, respectively, with bandwidth as the difference between these two frequencies). The number of inflection points were derived from peak frequency contours in Raven, which tracks modulations in frequency of a signal over time.

Preliminary analyses indicated that adult and yearling male songs did not differ significantly (Appendices 3, 4, 5), so we grouped these for the remainder of the tests. We assessed correlations among acoustic variables and removed one within each pair of highly correlated variables with a Pearson correlation value of $R = 0.8$ or greater (Appendix 6; Asuero et al. 2006). There was one pair of highly correlated variables: full song duration and number of syllables ($R = 0.95$). Therefore, we excluded number of syllables from the final analyses (Andersen and Bro 2010). Full song duration and syllable rate were log-transformed, and a squared transformation was applied to minimum frequency and percent pause to meet the assumptions of homoscedasticity (Jan et al. 2014). One outlier at the individual level and two at the song level were removed from minimum frequency to ensure a normal distribution. To ensure that the transformed data reflected accurate patterns, each of these four variables were also tested using the non-parametric *nparLD* package in R (v.2.1), which approximates an ANOVA-type statistic using the *F*-distribution with adjusted degrees of freedom (see Noguchi et al. 2012 for more details; Appendix 7). In order to confirm relationships between our variables, we also performed a principal components analysis (PCA) using the individual-level dataset; this confirmed the correlations between pairs of variables and further supported our removal of the number of syllables variable (Appendix 8).

To test for individual and sex differences in acoustic characteristics, we performed a repeated measures ANOVA using the song-level dataset, with individual nested within sex (to account for within-sex variation) for the three songs from each individual. Post-hoc Bonferroni corrections, traditional and sequential (Holm 1979), were calculated to account for multiple comparisons.

We also quantified sex-specific variance for each variable of the individual-level dataset using coefficients of variation because acoustic traits under strong selection are more likely to show less variance (Barton 1990, Nowicki et al. 2001, Reinhold 2011, but see Houle 1992). All statistics were computed using R, Version 3.6.3 (R Core Team 2021) or IBM SPSS Statistics for Windows, Version 26.0.

• RESULTS •

Song rates

Male Orchard Orioles sang significantly more often than females; on average, males sang 2.35 ± 2.43 songs/min, while females sang 0.14 ± 0.35 songs/min (Fig. 2; $p < 0.0001$, $T = -4.68$). Of the 27 females for which we have rate observations, we observed 16 of them singing (59%). Of the 48 males for which we have rate observations, we observed 43 of them singing (90%). There was no significant effect of time of year on song rate for either sex: linear regression models to investigate the relationship between song rate and time of year resulted in R^2 values of 0.035 and 0.0009 for females and males, respectively. Quadratic models resulted in R^2 values of 0.0418 and 0.0014 for females and males (Appendix 4).

Acoustic measurements

After Bonferroni correction, the nested ANOVA determined that female and male Orchard Oriole songs differed significantly for five of the eight variables measured (Tables 1, 2; Appendix 9; Figs. 1, 3). Female songs were shorter ($F_{1,88}$ = 744.00, $p < 0.0001$), had lower maximum frequency ($F_{1,88}$ = 13.49, $p = 0.0035$), and smaller bandwidth ($F_{1,88}$ = 38.88, $p < 0.0001$). Male songs had shorter syllables ($F_{1,88}$ = 8.21, $p = 0.0435$) and a greater proportion of pause ($F_{1,88}$ = 16.98, $p = 0.0007$). For the remaining three variables, females and males did not differ significantly.

Females showed greater variation for all eight variables, as calculated by an average of each individual's coefficient of variation (Table 3). The coefficient of variation is calculated using the mean, which removes any effect of scaling differences between the songs of each sex.

• DISCUSSION •

Female Orchard Orioles sang often throughout the breeding season, and their songs were structurally distinct from those of males for five of the eight variables we investigated. Female songs were easily distinguishable by ear in the field, reinforcing these data.

Our study presents the first formal documentation of female song in Orchard Orioles. Female song in this species had rarely been recorded or described previously. We did not notice females singing in our first five years of fieldwork on this species, and there are only brief mentions of potential female song in the literature. For example, Enstrom (1992) stated that females sang a single short song phrase that was recognizably distinct from male song and was produced much less frequently. In the Birds of the World species account there is only one short sentence about this possibility: “Females may sing occasionally” (Bent 1958, as cited in Scharf and Kren 2020). Even people who have studied the species extensively have expressed skepticism at the suggestion that females sing regularly (W. C. Scharf, *personal communication*).

In this species, delayed plumage maturation likely contributed to this lack of reporting on female song. Female Orchard Orioles have uniform olive-green plumage; yearling males look similar but have black feathers on their face and throat. In the past, researchers may have attributed female song to yearling males that they did not get a good view of, or that had little black coloration. In our case, we had excellent views of each of the females we recorded and observed many of the females that sang performing female-specific behaviors, including nest-building and incubation.

Typically, definitions of bird song distinguish long and complex songs from shorter and simpler calls (Catchpole and Slater 2008). We defined songs based on acoustic structure and context. Operationally, we defined song as vocalizations containing more than three syllables; some birds were observed producing song fragments with fewer syllables, but these were usually produced by males immediately before singing a full song, or in agonistic contexts (Moyer, *personal communication*). Songs were only produced by adult birds, not by fledglings, which are only reported to make contact calls or begging calls (Jaramillo and Burke 1999). Additionally, the songs performed by females clearly differ from other types of simple vocalizations used by Orchard Orioles including chucks, chips, chatters, and one-note whistles (Appendix 10; Jaramillo and Burke 1999, Sturge et al. 2016).

Song rates

Females sang at significantly lower rates than males throughout the breeding season. Like males, females produced song well into June, and neither linear nor quadratic models testing the effect of time of year on song rate explained the variation in song rate found in the data. Our speculation that females might mostly sing upon returning to the breeding grounds proved unfounded, as statistically there was no evidence of a decline in song production as the breeding season progressed (similar to Cain and Langmore 2015). Thus, our hypothesis that female song in Orchard Orioles mainly functions in mate attraction was not supported. Alternative hypotheses, such as territory defense or pair-bond maintenance, will be directly tested in the future using playback experiments.

Acoustic measurements

We found that female and male Orchard Oriole songs were statistically different for five of the eight acoustic variables measured: full song duration, syllable duration, maximum frequency, bandwidth, and percent pause (Tables 1, 2). These results provide strong evidence that female song elements in Orchard Orioles are structurally distinct from male song elements. Many previous studies in temperate species have concluded that female song is structurally simpler than male song (Arcese et al. 1988, Baptista et al. 1993, Price et al. 2009, but see Rose et al. 2018, Kornreich et al. 2021). Before we can begin to determine how or if these sex-specific differences are relevant to Orchard Orioles, it will be important to investigate if males and females can even perceive sex-specific differences in vocalizations.

Song function

Perhaps as a result of the historical bias toward studying males in a sexual selection context, rarely observed elaborate traits have sometimes been assumed to be nonfunctional, a vestigial side effect of selection acting on males. Darwin wrote that elaborate female traits were maintained through inheritance of male ornamentations, which he proposed evolved primarily for mate attraction and male-male status signaling (Darwin 1871, Andersson and Iwasa 1996, Tobias et al. 2011). Other early studies proposed that female song is a nonfunctional, short-term behavior resulting from hormonal shifts such as increased androgen (Kern and King 1972, also see Langmore 1998). In Chestnut-sided Warblers (*Setophaga pensylvanica*), one study concluded that female song was likely a non-functional trait because it was produced so rarely (Byers and King 2000). However, only 5% of the Chestnut-sided Warbler females sang in that study, compared with 50% of the Orchard Oriole females we observed. This is a conservative estimate given our strict conditions required to include songs in the acoustic analysis data, because only individuals directly observed singing were included. In other words, if a bird was singing slightly obscured in a tree, even if a female was known to be there, we did not include these songs in the acoustic analysis. Furthermore, to maximize sample size, rate observations were stopped after 10 minutes in order to record new individuals. Thus, it is likely that a much larger proportion of females sang than reported.

In some cases, however, infrequent communication signals can also carry complex information (Hauser and Nelson 1991, Wilkins et al. 2020). For example, female Canyon Wrens (*Catherpes mexicanus*) sing significantly less often than males, but female song in this species has been clearly documented to function for resource and territory defense (Hathcock and Benedict 2018). Female song in Song Sparrows (*Melospiza melodia*) has been referred to as a "rare, but normal aspect of female behavior" and is thought to function primarily in female-female conflict (Arcese et al. 1988:49). Male Eastern Bluebirds sing 20 times more frequently than females (Rose et al. 2018), but female song has been shown to function for pair-bond maintenance (Rose et al. 2019). Many important behaviors, such as egg laying or forming long-term pair bonds, can occur quite rarely in the life of an organism but are clearly functional.

Thus, our result documenting that females sing less often than males in Orchard Orioles does not necessarily indicate a lack of adaptive utility for female song. Given that female songs have different acoustic structure from male songs, vocalizations in this species could potentially serve as a sexual identity signal or have distinct functions for each sex (Riebel et al. 2019).

Evolutionary implications

Differing selection pressures on each sex may lead to evolutionary changes in acoustic production. For instance, there are many examples of passerine males with higher frequency songs obtaining greater mating success, as measured by overall male quality, increased female response, and/or higher rates of paternity (Ratcliffe and Otter 1996, Christie et al. 2004, Byers 2007, Cardoso et al. 2007, Ripmeester et al. 2007). In contrast, frequent female song may actually carry fitness costs: in Superb Fairywrens (*Malurus cyaneus*), females who sang more often were more likely to lose eggs to nest predators, likely by alerting predators to their location (Kleindorfer et al. 2016, but see Odom et al. 2021).

Sex-specific vocalizations may also assist in sex identification during territory defense. For example, Baptista et al. (1993) suggested that male White-crowned Sparrows were able to distinguish between the songs of females and territorial males by recognizing the shorter, simpler female songs as non-threatening. Likewise, male Red-winged Blackbirds (*Agelaius phoeniceus*) recognize the vocalizations of their mates, whereas females are apparently unable to tell other females apart (Beletsky 1983a, 1983b).

As mentioned, many scientists have historically proposed that elaborate female traits were only maintained via genetic correlation with males, resulting in females maintaining male-like traits without any fitness benefits (Darwin 1871, Muma and Weatherhead 1989, Amundsen 1999). In order to investigate sex-specific selection pressures on acoustic traits, we compared variances in male and female songs to determine if females had significantly greater variance, which would be expected if female song was subject to relaxed selection pressure (Barton 1990, Reinhold 2011). Females displayed greater variation for all eight variables (Table 3).

However, recent studies examining selection pressures in songbirds have suggested that greater variability in female songs may be selected for if female vocal signals are more limited in diversity and duration (Collins et al. 2009, Wilkins et al. 2020). Variation in vocal behavior may also reflect variation in competitive interactions (Cain and Ketterson 2012, Cain et al. 2015, Austin et al. 2021, also see Tobin et al. 2019). Additionally, if female songs are important for sexual or individual identity, increased variation might be expected (Sandoval and Escalante 2011, Hahn et al. 2013). Social selection has also been shown to play an important role in the evolution of elaborate advertisement traits in both males and females, as many behaviors function in social contexts outside of mating and the breeding season (Tobias et al. 2011, Lyon and Montgomerie 2012). More research into the sex-specific selection pressures that Orchard Orioles face is needed to determine how or if the variability of female songs affects female fitness.

This text was first published in the online *Journal of Field Ornithology* 93(1):3, in 2022, with the title “Female song is structurally different from male song in Orchard Orioles, a temperate-breeding songbird with delayed plumage maturation.” Find the full text with appendixes, literature list, and tables via:

https://journal.afonet.org/vol93/iss1/art3/.

Chapter 3:

Folklores and Futures

(76)

(77)

(78)

(83)

(84)

(85)

(89)

(90)

(91)

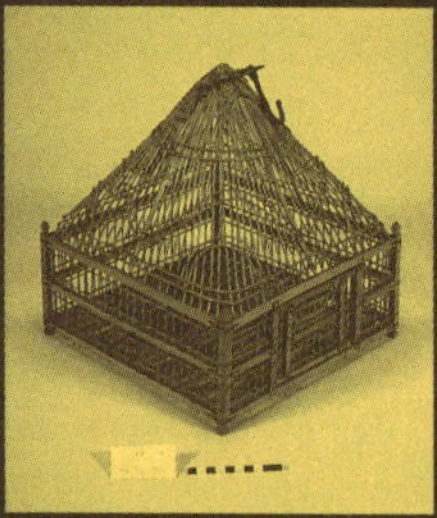

(95)

(96)

(97)

(98)

(80)

(82)

(81)

(88)

(87)

(93)

(94)

(100)

(101)

(102)

(77)

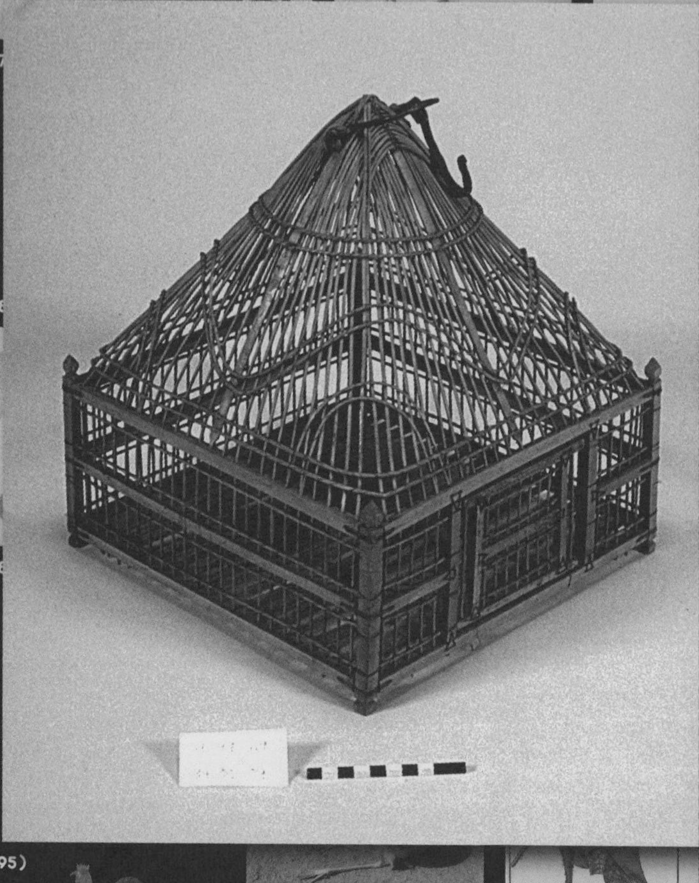

(95)

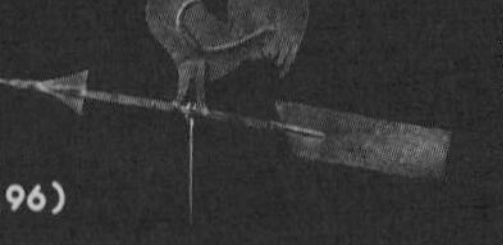

(96)

(97)

(98)

(82)

(88)

(100)

(102)

(76)

(77)

(78)

(96)

(97)

(98)

(80)

(82)

(88)

(100)

(102)

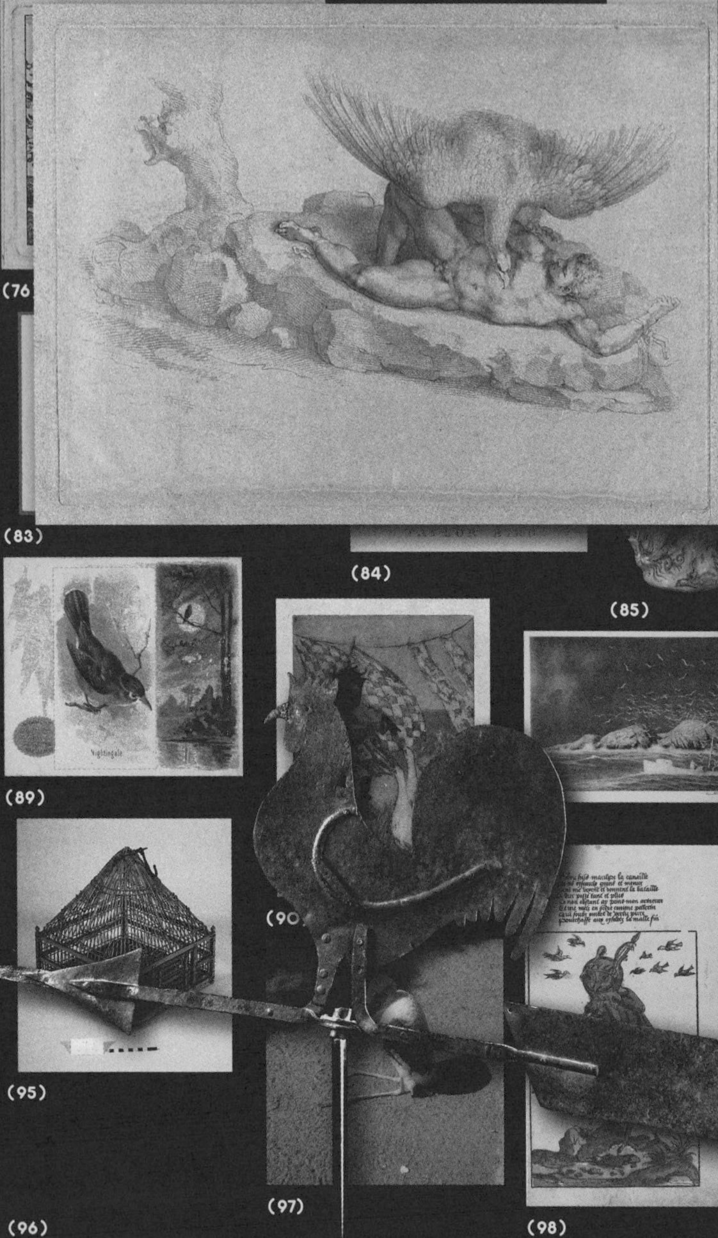
(76)
(83)
(84)
(85)
Nightingale
(89)
(90)
(95)
(96)
(97)
(98)

(100) (102)

(7

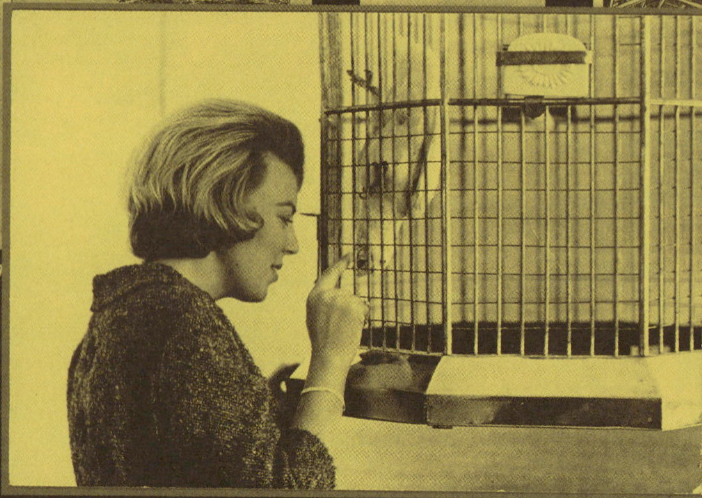

(83)

(84)

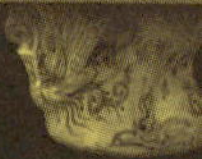

(85)

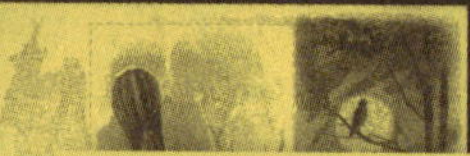

(89

(95

(96)

(97)

(98)

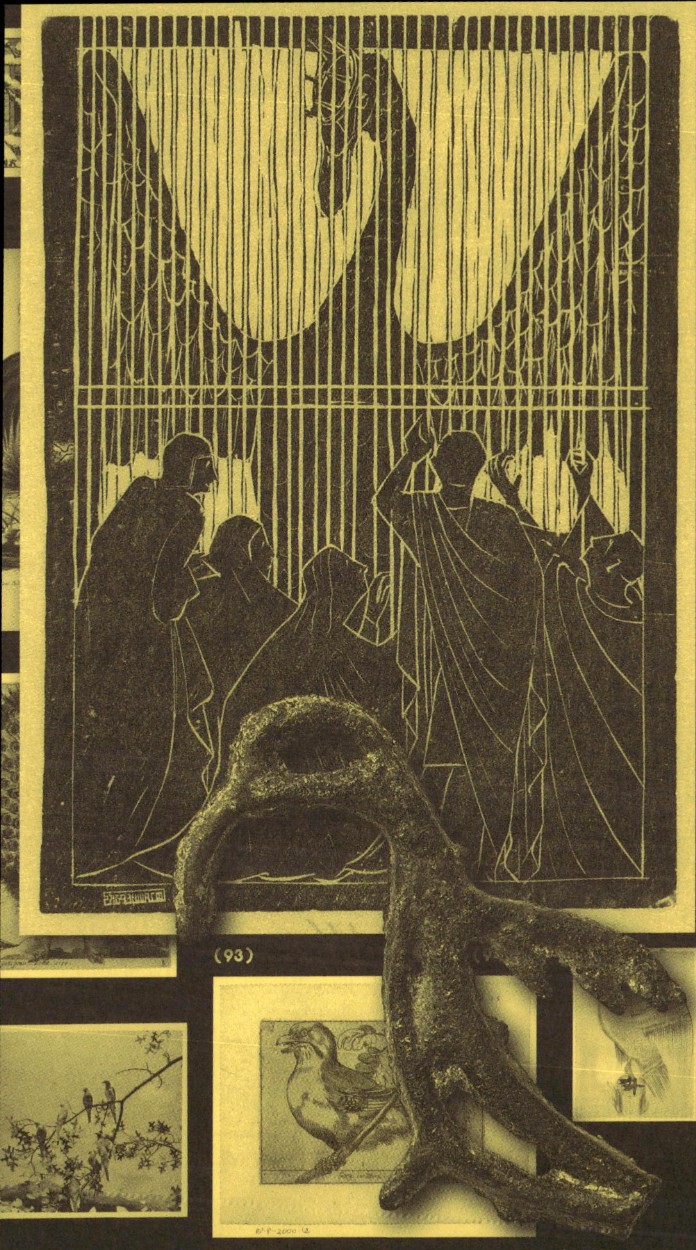

(93)

(100)

(102)

Hibou ſuis maulgre la canaille
De ces oyſeaulx grans et menuz
Qui me vevent et donnent la bataille
A leur poſte tant et plus
Ce non obſtant ay prins mon armeure
Et me metz en ordre comme pellerin
Qui ſoubz ombre de vertu pure
Pourchaſſe aux oyſeletz la malle fin

JOLLY OLE

TALLIO
BY IGNACE CAMI

As I sit here in my old olive green armchair, a coffee cup between both hands to ward off the morning chill, I watch the little birds outside my window. Their liveliness usually delights me, but today they seem agitated, ill at ease, and I'm not sure why. Then I hear it: *KROOAAACK*. The sound pierces through the silence. The tits and finches disappear into the foliage and the blackbird scatters toward the undergrowth. Again: *SQUAAAAA*. I know that voice. *KRROAAK* and the back of my head starts itching. *CROOAAA* as I remember. My fingers slide towards the—*SQUUUAAAC*—itch, until—*KRAAAACCCK*—I reach the tingling scar, and I'm transported to a time when this bird changed everything.

The stage: art school. A place that always seemed to escape reality, it had been my habitat for a few years by then. Students in full steam punk-attire; crusty weed smokers moving around in thick clouds; philosophy adepts swearing by Nietzsche, vintage glasses, and pipe smoking; that one perpetually drunk beatnik teacher—those were just some of the characters who inhabited my daily life. You could not imagine a stranger place.

After one of several endless and painfully grounding summer holidays, I returned to art school full of longing. I was keen to get back and dive into its madness. What would it have in store for me this year? What magic would await me? Would it surpass last year's werewolf episode, where a senior student, over the course of several weeks, took to biting the teachers and urinating all over the hallways in the name of performance art? Little did I know that the answer was in fact yes. This particular year, the magic would come in the form of a noisy, brownish-pink bird with black and white lining, a silly-looking crest, a black mustache, and a gorgeous splash of cerulean blue embroidered on its wings. He was a Eurasian jay, and his name was Tallio.

Tallio was obsessed with Ina, a fellow student who got the bird from her uncle. The uncle had found Tallio as a helpless hatchling during a hike, brought him home and hand fed him for a few days. He decided to give Tallio to Ina, who took an interest in the little bird, as a project for the long summer months. They immediately hit it off. In his strange way, Tallio was the first one to make Ina feel like she wasn't invisible. He seemed to be the perfect companion—a creature who was the exact opposite of her: loud, bubbly, and colorful. As Tallio grew stronger, he began to sit on Ina's shoulder. All day long she would use her sweet and high-pitched voice to talk to him, his ugly little head tilting and nodding in delight. By the time Tallio's plumage filled out they had become inseparable. He would jump and flutter, crest raised like a punk, and gobble weird and throaty hoots, occasionally pierced by a chilling *SQUAAAAACK*. It was Tallio's presence—his way of listening, of chirping in response to her thoughts—that seemed to ignite something in Ina, something she hadn't fully tapped into before.

In previous years, Ina had been a slightly underperforming student with few to no friends. She came off as an introverted soul, quiet and reserved. For a long time she was completely overlooked. In fact, it seemed to be her natural state—content remaining at the edges of conversations, observing rather than speaking. But now, with the bird on her shoulder and a crowd growing around her, that was all about to change.

Ina's bond with Tallio grew deeply intimate. You might even call it spiritual. She talked to him constantly, pouring out her thoughts, ideas, and frustrations to the bird. The more time they spent together, the more it became clear that he understood what she was saying. When she woke up one morning and found Tallio staring at her from the windowsill, she wasn't even shocked when he said "Good morning." It was only later that day that she realized that the bird had talked to her. That's how natural it felt. The young bird was always by her side and now it could finally have its say. The thought was thrilling. Later that day, as she was sketching, Tallio began to feed her insights, stories, and strange visions that she eagerly put to paper. This went on for weeks, Tallio and Ina conversing, sharing and confiding together as he sat by her side. She had never felt more productive.

One afternoon, while Ina was busy preparing a canvas, Tallio did something that would have a profound impact on her work. Having gained strength each day of their companionship and his feathers finally having filled out, he unexpectedly took flight. He kept zipping across the room, his wings flapping in a frenzy. Suddenly, he made a sharp turn mid-air, his wings brushing against the wet canvas in a blur of color. Ina gasped as a trail of cerulean blue streaked across the surface, leaving a mark more vivid than any brushstroke ever could. Something clicked inside her. She didn't have to force her creativity. She could let go. She could let the energy of the bird, the wind, the world outside, shape the flow of the paint.

Soon after, her art blossomed. She could feel the energy of Tallio in every line she drew. A few weeks into the semester, Ina started producing rapidly, churning out painting after painting. The works were strangely intriguing and looked like nothing the students had ever seen before. For a long time, her work had been perceived as mediocre at best, materializing as uninspired portraits filled with technical flaws, lacking the spark that so many artists yearn for—the feeling, the image of transcending. But last year's portraits had now made way for an absolutely mesmerizing abstraction. Everything was different; the brush strokes forged lines so free they seemed to float over the canvas, punctuated by powerful explosions of color so deep you were at risk of plunging in.

Ina had become a sensation overnight. Her paintings were the talk of the school. Whenever she walked through the hallways, Tallio chirping on her shoulder, students would flock around her. They would complement Ina on her latest painting and pet Tallio on his head. Sometimes they brought nuts to feed him, for which he made his appreciation known by hooting frantically, a sound that became widely imitated throughout the school. Their interaction was friendly, but they never heard him speak. That gift he saved for Ina, whom he adored. He was her muse, she his idol, and together they were unstoppable.

Unstoppable too, though, was the inevitable rise of gossip. At first, there was disbelief. Who would have thought Ina would ever make anything this good? How did she end up going with abstraction? Where did she learn how to do that? Soon after, jealousy reared its ugly head. Some teachers began criticizing her paintings, calling them savage and lazy. They declared abstraction redundant and passé, but Ina couldn't care less. She was having the time of her life, painting one canvas after another, flowing. As the semester went on, her paintings only became more electrifying. It was no longer just the intensity of their color that drew attention; it was their sense of breathing—as though each piece had a real life of its own. Ina often explained that the paintings were "singing" to her, that it was Tallio's voice she heard in them. His unrelenting presence had become ingrained in every piece. When she took a step back—away from her easels, eyes heavy with sleep—she would see the flicker of Tallio's silhouette in the corners of the room. Even when he was quiet, he was always there, his presence persistent in every piece of art she made.

It was not long before rumors of eerie giggles and strange voices coming from Ina's studio caught on. True, she hardly ever went out, but now she was said to be talking to someone in there all the time. I remember hearing it once myself. I was standing in the courtyard, right under her window, and I could hear two distinct voices. One was clearly Ina's, but the other sounded off in some way, so I decided to climb one of the slender nearby trees and have a look. I chose the tree closest to Ina's studio. As I climbed higher, the branches became thin. Gradually, the voices grew more audible, but they were still largely obstructed by the loud and slapping sound of Ina's brushes against the canvas. So I kept climbing, and just as I could see the top of Ina's head over the edge of the windowsill, Tallio landed in front of me with a loud shriek. As I jerked backwards, the branch beneath my feet gave way, and the next thing I knew I was lying face up on the courtyard floor, with Ina looking down at me from her window. She screamed in horror as I felt a warm liquid spreading beneath my head. As I slipped in and out of consciousness, Tallio landed on my chest. He turned his left eye towards me to take a closer look, and I could have sworn he asked me if I was all right. When I woke up in the hospital, I learned that I had been treated for a head injury. The recipient of eighteen swirly stitches, I was sent home for two weeks of rest, to recover from a severe concussion.

I found healing to be a slow and boring process, and I was eager to get back to art school. Tallio and Ina were constantly on my mind. When my weeks of recovery were finally over, I returned to find a shift in the school's dynamic. During my absence, a new teacher had been inaugurated—a big deal in the arts, a man with prestigious accolades. Even his name, Kasper Roose, transmitted authority and promise. Roose, as we would call him, was a self-proclaimed "expert in technique." Regrettably, though, he was nothing like the experimental artist we had hoped him to be, our hopes based on the artworks we had seen in his books. In fact, he turned out to be a bitter man. Roose's own art had fallen into malaise—his works, once filled with bold experimentation, were now just murky blobs of paint, lifeless and stagnant. If his failure was palpable, his bitterness was even more so. He was like a withering flower looking to others for substance, sucking the life out of any room he entered. His sullen character quickly weighed down on the spirit of the students, and whenever he would make his round of studio visits he would leave them drained and confused.

Roose resented students who showed promise, especially those who seemed to effortlessly achieve what he had been failing to attain. His own success was fleeting as he struggled to stay authentic. Of course Ina, shining brighter than ever, was the peak of his frustration. He had heard the stories of Tallio, the musing bird, but it was Ina herself who told him that the bird would speak to her, giving her access to a raw and chaotic creativity. The lightness with which she shared this made him green with envy. Roose took an unsettling interest in Ina's process and often addressed Tallio directly during his visits, but the bird gave him nothing. He only had eyes for Ina. Roose would relentlessly question Ina on how this relationship worked. After all, nobody had ever heard Tallio say anything, really, let alone seen it make art. It was just a loud and quirky bird. Social, yes, but conscious? Intelligent? Ina insisted: the bird was the key, and his words unlocked her raw potential. "You are not a puppet," Roose would repeat, "you don't need a bird to paint." But he knew he was telling this to himself more than anything, as in Tallio's eyes he saw something that he could not fully dismiss. Could it really be true? All that talent, from a jay? If it really was a feather-wrapped generator of creativity, then why should it be hers? What made her so special? Had he not suffered for his art? Was he not deserving? He had to know. For weeks the corrosiveness of his spite had been eating away at his principles, which eventually paved the way for his despicable plan.

For days Roose lingered in the hallways closest to Ina's studio, waiting for an opportunity to arise. Ina rarely left, though, and when she did, Tallio would be sitting on her shoulder, chirping away. But spite is a powerful force, and Roose endured. One afternoon, Ina finally left her studio to attend a mandatory lecture in the school's auditorium—not a good place to bring a noisy bird. Without hesitation, Roose burst into the room and grabbed Tallio firmly. The startled bird would soon revamp his dying art practice. He threw Tallio into a small iron birdcage and covered it with a piece of fabric. Roose, high on the promise of success, barged around the room and, confronted with Ina's mesmerizing paintings, lashed out before leaving. That evening the school was shaken by a piercing cry. Ina had returned to find Tallio missing, and a large and looming slit scarring her latest painting.

When I heard the news I helped Ina look for Tallio, but to no avail. He seemed to have disappeared from the face of the earth. After a long day of frantic searching, Ina sat in front of the darkness gaping down the middle of her canvas and wailed. She seemed inconsolable. How quickly things had changed. I suggested that, in an effort to distract herself, she could try painting again, but the colors, once so alive, now felt flat. Her brushstrokes had become lifeless. Her thoughts inevitably returned to Tallio. Where had he gone? Why hadn't he come back? Other students steered clear of the scene, not knowing what to say or do. Even though I was there, she was alone, for the first time in months, and it was tearing her apart.

On the other side of town, Roose stared at the now-caged jay. The bird was kept under full and sterile scrutiny in Roose's fluorescently lit studio. The workspace was neat to the point of obsession. His desk, covered with meticulously arranged paint tubes, brushes, and a stack of books on technique, realism, and traditional methods, was aching to be used. And the bird would be instrumental, for it would break Roose's stagnation and restart his career. But Tallio did not speak. Much to his frustration, Roose's pleas were only answered with coarse caws. The space was full of well-maintained easels, each holding a large, white canvas. On the bottom of each canvas he had already placed his signature—his last name materializing in a bold, arrogant flourish, as if to remind the viewer of his once revered status. Now and again he retreated to a worn leather chair in the corner of the studio, its surface cracked from years of use, where he would rest and brood.

On the table next to it, his own artist monograph *Blooming* mocked his current inertia, the sight of it pushing Roose back towards his easels, and to Tallio. Standing before the cage, Roose's eyes locked with the bird's. It was staring at him—not in anger, but with a deep, unflinching calm. Its vibrant blue feathers seemed to glow softly under the harsh lights. Roose reached a trembling hand toward the cage, and for the first time since he had taken the bird, he felt a flicker of guilt in his chest. His fingers brushed against the bars and, replacing his usual rush of frustration, he felt something else—something unexpected.

It was as though Tallio was looking *into* him; past the bitterness, past the jealousy, past the self-pity, and settling on something else. The bird's gaze was steady, unwavering, as though it understood something Roose couldn't yet grasp. The resentment in Roose's chest flared momentarily, but then the bird blinked, and for a split second something softened within him. The room felt warmer somehow, as rays of sunlight cut beautifully through the studio windows. Roose's chest rose and fell slowly, as if he were catching his breath. To his own shock, he found himself murmuring.

"Can't you help me?"

His voice was rough, barely a whisper, but in the stillness of the room, it felt like the most honest thing he had said in years. Still, Tallio did not respond, not in the way Roose had hoped. For the first time, Roose saw the bird for what it was: a creature of pure instinct, untainted by ego or expectation, flying freely with no need to justify itself. The weight of this realization hit him hard. He was hot with envy. He snapped, flooding the room with insults and threats, poking Tallio through the bars and rattling the cage frantically. Roose bolted out of the studio screaming, but the bird remained silent. All he cared for was Ina.

When Roose had finally calmed down and returned to his studio, he couldn't believe his eyes. Tallio had broken out of the cage and shat on every single canvas in sight. Vile streaks of color oozed down each primed surface tauntingly. He grabbed the first thing he could get his hands on, a book on oil painting, flung it at the bird, and missed. Roose grabbed another book, this time on realism, and threw it at Tallio. Once again he missed the jay, and in his rage had thrown the book straight through his studio window instead, shattering the glass. They both realized this was it, it was now or never. Roose and Tallio both rushed towards the opening, and as it became clear to Roose that he couldn't outrun the flying bird, he grabbed another book and flung it towards the broken window, envisioning its trajectory clearly in his mind. The book—Roose's own artist monograph—flew across his untouched desk, past the poop-covered paintings, and loudly smacked the bird against the windowsill. As the window frames rattled, both book and bird fell to the ground, lifeless. Roose ran towards the place of impact as though in a hurry to halt the damage, to fix the situation, but deep down he already knew what he had done.

That night, sitting in her studio, Ina heard a familiar fluttering of wings. She looked up hoping to find Tallio, but instead was greeted by a wood pigeon in the tree outside, adjusting its wings one last time before it would settle down into sleep. Ina stood slowly and stared into the darkness outside. The pigeon just sat there, still, and the heavy silence between them lingered. "Tallio?" The bird tilted its head and blinked its eyes before pulling in its neck and dozing off. "You're just a bird." Ina turned around and stared at the gouge in her painting, darkness still exuding from within it, and she heard his voice. "It's never just a bird." Tears rolled over her cheeks. Her beloved bird had gone, but she could feel him still.

The next morning, as rays of sunlight peeked through the small tree in front of her studio window, Ina awoke, stretched, and changed. The weight of not knowing had fallen from her shoulders. She was freed from the need to understand or explain. Driven from within, she grabbed a needle and thread and began to sew. Stitch by stitch she started mending the wounds in the canvas, and she did not stop until her painting was whole. It looked better than it ever had before. These rough stitches of thread closing off the darkness would become the best-known signature of her soon to be sought-after work. This very first one, though, she gave to me.

This painting sits proudly on my living room wall, alive with the memories of all that transpired back in art school. It's more than just an object now—it's a story, a transformation, a lingering presence that's been woven into the threads of my life itself. The painting is one of her best works from this time, later titled *Echoes in the Current*. Though the canvas is enormous, it isn't the scale that takes your breath away. It's the vibration in the air when you stand before it. The sense that it's alive, humming with a pulse of its own. Every day I look at it, and every day I discover new things in it. It is an ethereal exploration—a swirling symphony of shapes, colors, and tactile patterns. It feels less like a painting and more like a moment, frozen in time, as though Ina captured the ebb and flow of a vast, unseen current of energy and conveyed it through the chaotic fluidity of brushstrokes. The colors are intense: electric blues, rich purples, and glowing oranges melt together in streaks of motion, like currents of water pulling in different directions. There are deep, deep crimsons that morph into pale, almost iridescent pinks, its tones in perpetual flux.

What stand out most, however, are the soft, organic shapes embedded in the work—sinuous, flowing arcs that suggest the outline of something but never quite define it. Something just beneath the surface, unseen but powerfully felt. There's a sense of freedom, of swift and fluid movement that ripples outward from the center. It's almost like the painting itself is a living, breathing thing, and yet there is stillness at its core, held together by those bold stitches. Rough threads, confidently applied, as if to say: I will overcome. The sight of it still makes the back of my head itch.

In the lower-right corner, the paint seems to break apart; the edges of the swirling currents become jagged and fractured, pulled outward by some kind of force but still tethered together by an invisible string. This section is somber and hints at a darkness that is repressed by the intense light and energy of the rest of the composition. But if you look closely, you'll see a few cerulean blue feathers, half encased in paint. That's the part I like the most.

I still see Ina from time to time, but really we drifted apart. Her art—a rich combination of textiles and paint—has matured, and is shown all around the world. Everywhere she exhibits, her work is praised. And everywhere she exhibits, she keeps looking up. She scans the trees and the sky in hope that one day, somewhere, she'll find her Tallio.

HYBRID BEINGS

ENTANGLED LOVE IN AN ENDANGERED WORLD

BY MANJOT KAUR

Hybrid ***Beings*** *is an artistic and conceptual exploration that proposes an interspecies spectrum of entangled existence.* ***Hybrid Beings*** *is a series of paintings rooted in fabulation and speculative postulation, depicting post-human species that extend an invitation to re-imagine the connections between species, identity, and ecology. It challenges patriarchy and gender bias, and offers a vision where the boundaries between human and more-than-human dissolve. I initiated* ***Hybrid Beings*** *in 2020, as an exercise in worldbuilding in my visual art practice through contemporary miniature paintings.*

As a series of artworks, ***Hybrid Beings*** *visually narrates concerns of the nature-culture duality, the sixth mass extinction, habitat destruction, loss of biodiversity, and the imbalance between human and non-human life.* ***Hybrid Beings*** *proposes a speculative world where notions of humanity, identity, hierarchy, patriarchy, and heteronormativity are disrupted.*

As a text, **Hybrid Beings** seeks to reach viewers and audiences who are left to wonder what these paintings mean, grappling with their contexts, symbols, and the layers of possible interpretation. This text is offered as a key to address the artworks, their conceptual framework, and the embedded questions of loss, repair, kinship, and reciprocity. To expound on these ideas and to narrate **Hybrid Beings** as not a series but a speculative species, this composition of writing is further divided into the sub-categories of storytelling, kinship, the post-queer, the post-human, the uncanny, dismantling hierarchies, and repair. This sub-categorization elaborates in detail on who **Hybrid Beings** are as a species, unpacking how they conceptually propose alternatives to both our present and future.

Before proceeding, it is important to first explain the conceptual premise on which **Hybrid Beings** as a series of paintings is built, so that the sub-categories discussed later can be more effectively engaged with.

The women and stories in **Hybrid Beings** are inspired by Ashta-Nayika, a collective term for eight heroines in various states of love, as classified by Bharata.[1] The heroines each embody a distinct emotional state, mood, and position in relation to their lover, ranging from longing to betrayal to devotion. Abhisandhita Nayika has been estranged from her lover after a fight, and is full of sorrow and regret for her behavior. Bold and brave, Abhisarika Nayika is the daring one who goes out into the forest to seek her lover.

Khandita *Nayika is enraged after being betrayed by her lover, who, despite promising to meet, spends the night with another.* **Proshitabhartruka Nayika** *humbly waits for the return of her husband, who is out on business and has not returned on time.* **Swadninapatika Nayika** *is in control; her lover is loyal and obeys her, subjugated by her immense love.* **Utka Nayika** *is distressed and pining as she waits anxiously for her absent lover. Also waiting for reunion is* **Vasakasajja Nayika**, *who is deeply involved with her lover, has a desire to lead a life with him, and wants to express this to him.* **Vipralabdha Nayika** *is neglected by her lover, who promises her to meet but never shows. Feeling deceived, she waits for him the whole night in vain.*

---

1 In *Natya Shastra*, Bharat Muni's Sanskrit treatise on performing arts.

Traditionally articulated through classical art forms — sculpture, poetry, music, dance, and painting — in the **Hybrid Beings** *series the concept of* **Ashta-Nayika** *maintains its familiar painterly surface while reimagined within the framework of hybridity to explore new possibilities of interspecies relationships.* **Hybrid Beings** *portrays women and birds in various romantic encounters, where love, grief, and desire become sites of ecological repair, aiming to decolonize the biased and anthropocentric thinking of human beings. The birds become the heroes, replacing the human male figures of the traditional* **Ashta-Nayika**, *and merge with the women, thus creating* **Hybrid Beings**. *The birds depicted are primarily endangered or extinct species from around the world, including the white-bellied heron,*

Bengal
florican, Jerdon's courser,
O'ahu 'akialoa, great Indian bustard,
ivory-billed woodpecker, common house sparrow,
piping plover, Sumatran ground cuckoo, common
kestrel, and yellow-breasted bunting, among others.

Grounded in posthumanism, animism, and feminist critique, Hybrid Beings asks: How do we mourn what we have lost? Where can we place our efforts to regain the biodiversity that teeters on the verge of extinction? How can we create a sustainable and respectful correlation between biodiversity and human progress to future-proof our planet for generations to come? How can we amplify the voices of the wild?

The term "Hybrid Beings" is used interchangeably in the text below, signifying a conceptual framework, the stories encompassed in the series, and, predominantly, a speculative species.

ON STORYTELLING

Hybrid Beings are visual allegories proposing solidarity among species and creating possibilities of entanglement with unexpected companions.

Hybrid Beings inhabit precarious, pristine, and fragile habitats.

Hybrid Beings are situated in eco-emotional landscapes, where the flora, mountains, soil, and water are wunderstood as sentient beings.

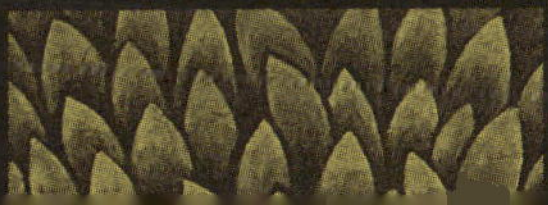

Flora embraces and encapsulates the Hybrid Beings.

Hybrid Beings weave vulnerable ecologies into a process of worldbuilding beyond the constraints of reality with an infusion of awe and wonder, and depict romantic interspecies encounters. The bird-headed heroine waits for the lover bird on a bed nestled in the mountains of clouds or within a hovering palanquin in a mangrove forest. Love letters are written on a flying bed, and the company of loved ones is enjoyed while resting on a couch of leaves and moss. At times, Hybrid Beings are full of grief, or surrounded and engulfed by a polymorphic tree, and at other times they play a serenade on the tree of mountains, or prepare for a tryst on the tree of clouds. Occasionally, they fall in love with trees and horses while gliding in space.

Operating across multiple time frames, Hybrid Beings embody a synchronous sense of the past, present, and future, experimenting with alternative relational forms. The stories disturb the operation of convention, and are polytemporal, with species egalitarianism at their core.

ON KINSHIP

Hybrid Beings stitch together improbable collaborations in a multispecies world, making way for species-fluid kinship. They symbolize the potential for harmonious and mutual interactions between species.

Hybrid Beings make odd kin with other species, and pose the question:

"What
will be my response to
ecology, if ecology is my beloved?"

Hybrid Beings invoke a philosophical, personal, romantic relationship with other beings that coexist(ed) on the globe with the human species. Hybrid Beings embody the lover in the body of the loved. Women take on the identity of their lover-birds and shed their human heads.

Hybrid Beings generate hope and care, cultivating the capacity to reimagine a future for the marginalized and silenced; birds, ecosystems, and land alike. They disrupt normative ideas about biological kinship to remember the forgotten notions of care and symbiosis.

Hybrid Beings are personal totems that propose philosophical romantic relationships between women and ecology as a way to mourn what is on the verge of being lost.

ON THE POST-QUEER

Hybrid Beings surpass conventional ontological categories. They suggest that gender is not fixed, but rather can be diverse and multifaceted, encouraging an inclusive understanding of identity.

Hybrid Beings transcend the traditional meaning of hybrids as monsters or tricksters.

Hybrid Beings emphasize the fluidity of gender and the diversity of experiences in the natural world.

Hybrid
Beings imply that identity
is an evolving and adaptable concept.

Hybrid Beings confront heteronormative and hegemonic conventions, envisioning a free world that transcends the confines of traditional categorizations of gender. Women in Hybrid Beings are sovereign, with the potency to choose who they are, who they love, and if and when to reproduce, without explaining the "whys" of their decisions or proving the correctness of their being.

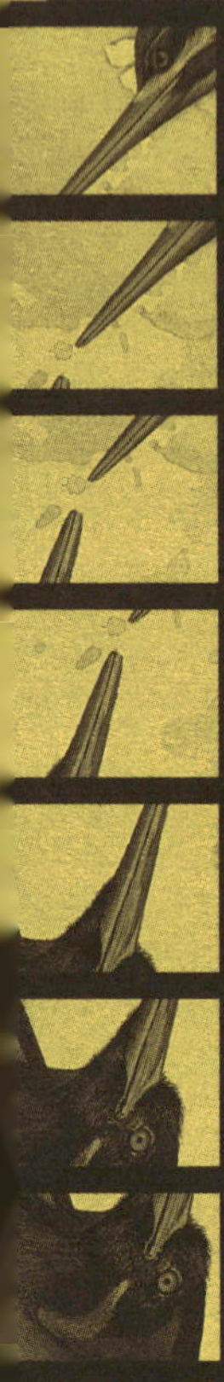

Hybrid Beings transform human beings into creaturely beings.

Hybrid Beings call for a holistic approach to understanding the ecological web of life, where identities and relationships intermingle, celebrating the diversity of romantic and ecological relationships.

ON THE POST-HUMAN

Hybrid Beings challenge the conventional idea of heroism rooted in patriarchy, and position birds, women, and the environment as protagonists.

Hybrid Beings reclaim power and agency for the birds and their habitats, and open windows of freedom for women to exist untethered.

Hybrid Beings grant women bodily autonomy and self-determination in reproductive choices, and release them

from the
tight weave of patriarchal,
religious, socio-political, and
economic binds.

Hybrid Beings form cross-stitched generations of not-yet-born and not-yet-hatched vulnerable and coevolving species.[2]

2 Donna J. Haraway, *Staying with the Trouble: Making Kin in the Chthulucene* (Duke University Press, 2016).

Hybrid Beings search for relationality, reciprocity, and mutuality within natural, ancestral, and human worlds. Rather than occupying the wild, Hybrid Beings inhabit it.

Hybrid Beings dissolve the boundaries of love, connection, and species hierarchy.

Hybrid Beings peer into different temporal dimensions, bridging the gaps between past, present, and future. They serve as a lens to meet future and past beings, future and past myths; those which have become extinct with evolution, and those that are yet to evolve either naturally or with the bio-geo-hacking of futuristic technology.

Hybrid Beings raise questions about the responsibilities and ethics of human society toward and within the natural world.

Hybrid Beings establish that intelligence,
once considered the domain of humans,
is shared across species.

Hybrid Beings accept that non-human beings exercise language, tools, and consciousness, as well as exhibit capabilities beyond human limits, such as echolocation, extraordinary vision, and powerful sensory perception, amongst others.

In light of this knowing and not knowing, perceiving and not perceiving, Hybrid Beings place humans in a continuum of evolution and intelligence with other species of flora and fauna, as well as abiotic aspects of the landscape.

Hybrid Beings cultivate the capacity to reimagine wealth and witness grasses, soils, mountains, trees, leaves, plants, fungi, vines, water, and moss as sentient beings.

Hybrid Beings are in a continuous process of evolving and becoming.

ON THE UNCANNY

Hybrid Beings open up possibilities for a post-queer and post-human world, where species move towards an uncanny kind of becoming.

Hybrid Beings embrace the peculiarity and obscurity inherent in the merging of identities.

Hybrid Beings challenge the viewer to confront the disturbing and navigate the discomfort of the unfamiliar.

They bring into being a sense of disquieting familiarity and strangeness simultaneously.

Hybrid Beings embody the recurrence of something long-forgotten and repressed. They become a reminder of our psychic past while at the same time pointing to a future dominated and supported by technology.

Hybrid Beings serve as a portal to a reality where the uncanny becomes a stimulus for symbiosis, transformation, and contemplation.

Hybrid Beings become a means to reimagine our relationship with the land and its myriad inhabitants.

ON DISMANTLING HIERARCHY

Hybrid Beings push back against the human-as-protagonist and move toward a thinking that eradicates the hierarchy of being to challenge the human/non-human binary.

Hybrid Beings serve as powerful reflections of a future where the barriers between human and non-human entities blur, advocating for an inclusive, interconnected, and uncertain evolution of life forms and identities.

Hybrid Beings dismantle the established hierarchies in social structures and encourage viewers to

reconsider
the rigid boundaries that
have traditionally separated humans
from the rest of the natural world. Hybrid
Beings imagine a world where humans are not the
apex species, but rather a part of a larger whole.

Hybrid Beings portray multispecies worlds where relationships between humans and non-human entities are not limited to exploitation or domination but can be characterized by love and grief.

Hybrid Beings present a profound existential inquiry, urging a reexamination of humanity's place within the intricate tapestry of existence.

Hybrid Beings make way for an ontological pursuit of what it means to be human, what it means to be non-human, and where these categories rupture and collide.

ON REPAIR

Hybrid Beings extend beyond the romantic relationships between women and birds and lay the foundation for a broader ecological imperative of working together with non-human animal and plant species to address environmental crises.

Hybrid Beings place endangered birds at the center of the narrative, emphasizing the urgency of conservation and the need to recuperate, restore, and protect their respective habitats.

Hybrid
Beings suggest that the
land equally belongs to other species as
much as it belongs to humans.

Hybrid Beings may or may not choose to reproduce, therefore they become a trope to negate capital and extractive economies.

Hybrid Beings invite reflection on the consequences of human actions and their impact on the fragile ecosystems of the planet, reminding us of the delicate balance within our environment.

Hybrid Beings emphasize the need to recognize and remember that both biotic and abiotic elements of the landscape are living beings, not mere objects to be trampled upon by humans.

Hybrid Beings find ways of existence and modes of happiness without treading on the well-being of others.

Hybrid Beings offer a profound invitation to envision a post-human reality, where the fusion of species and the fluidity of identity prompt a reconsideration of humanity's role within a biodiverse ecosystem. They raise numerous questions and ask how "we" as a species could commit to the rewilding of the land. Can we leave more space wild, to the wilderness? How can we find more ways of reciprocating with the land? How can we learn to listen to the earth, given that humans and plants communicate in entirely different modes and frequencies? How can scientific research and artificial intelligence help to make

visible the intelligence of other species in our daily life? Can we accept that land could own itself and should have rights like humans and corporations? How can judiciary and political assemblies, and other human-made power structures, acknowledge this intelligence, and make laws that uphold the life and ways of living of the more-than-human world?

This text, **Hybrid Beings**, *functions as a manifesto. It beckons towards a future where the wild is a gateway to understanding and embracing the sovereignty and complexities of diverse, healthy ecosystems.*

Hybrid Beings *are a reminder, "If grief can be a doorway to love, then let us all weep for the world we are breaking apart so we can love it back to wholeness again."* [3]

---

3 Robin Wall Kimmerer, *Braiding Sweetgrass: Indigenous Wisdom, Scientific Knowledge, and the Teachings of Plants* (Milkweed Editions, 2013).

THE ART OF DECEPTION
BY BRYONY DUNNE

The sun had just faded and the air felt clammy at the height of midsummer growth along the banks of the River Deel. In its brackish waters, which wind through the small village of Askeaton and below the ruins of an ancient Norman castle, a tall grey heron could be seen perching on one foot. I wondered whether this heron always came back to the same place to watch the waters fleeting between the mossy green rocks, and what stories, warnings, and fear the water brought with it as it passed the bird's watchful gaze.

***

It was the summer of 2023 and I had been invited to participate in an artist residency program in the rural village of Askeaton, located in County Limerick, Mid-West Ireland. For the two-week duration of the program, organized by Askeaton Contemporary Arts, artists are encouraged to carry out research that engages with the local community and surrounding rural environment. With no white cube gallery spaces in the village, artists present their accumulated research to the public on the final day of their residency in alternative spaces around the area. Sites available include town halls, pubs, car parks, petrol stations, hardware stores, and the river.

One of the works I produced in Askeaton consisted of three wooden bird decoy sculptures. With the help of a brave technical team, rigged out with sweeping brushes to balance themselves while crossing the river, we carefully placed the decoys in the polluted waters of the Deel, which flows through the center of the village. The birds took their position just in time for the arrival of a bus full of guests, here to take their tour of the artworks on display throughout the village.

This river is known as *An Daol* in the Irish language, which means beetle or worm, and is named as such because of its shape, which mimics a slithery worm, bending and carving its journey through the landscape. Along its banks that summer, peregrine falcons were nesting on top of one of the towers of the Norman castle. A little further upstream, the ruins of an old Hellfire Club—a 17th-century gentleman's club—can be seen from the water. This was one of only two clubs in Ireland where aristocratic members would converge to drink, discuss political and economic matters, and exchange their accounts of recent hunting and game jaunts in the surrounding countryside. Murky stories surface from its red brick walls; members were known to shapeshift from goat to man, man to goat, ram to man, man to ram, and so forth, whilst playing card games late into the dark winter nights.

Because of its tidal form, this river landscape naturally shifts and reshapes day after day; a flux that echoes the shifting nature of the stories and myths that emerge from this environment. One of the bird decoys positioned at the foot of these old stone ruins feature the head of a goose, while the others are taken from the symbolism and imagery of otherworldly beings, drawing on the connection between rivers and female deities in Irish mythology. These immortal beings were feared and respected, as rivers were considered to be sacred and vital systems bringing forth fertility and an abundance of life to the land that they journeyed through.

Technical team positioning bird decoys, River Deel, Askeaton, photo: Bryony Dunne.

Wooden bird decoys (work in progress) on the River Deel, Askeaton, photo: Sean Lynch.

During the British occupation of Ireland, the region of the Shannon Estuary and its tributary rivers flowing into the estuary—one being the River Deel—were highly prized by the landed gentry due to their exceptional hunting and fishing opportunities. At the time, power, status, and influence could be equated to the richness of the game found on one's property. Migratory and resident birds such as curlews, snipe, egrets, golden plovers, kingfishers, cormorants, brent geese, and whooper swans were to be found in abundance. Rivers were well stocked with wild Atlantic salmon and European eels, the latter now critically endangered. These lush lands, rivers, and estates functioned as places of leisure, used to entertain guests who often traveled long distances to enjoy the bountiful fauna and company. Many of the grand houses and castles of these sites can still be observed along the river banks.

Further up the River Deel, nitrates from farm and industrial waste and sewage from outdated treatment facilities creep their way into the briny waters. These substances contributed to the highly polluted state of the water quality that summer. Warning signs reminded passersby not to bathe in or drink the river, and I hoped that my wooden decoys would send a message too. Perhaps they would draw visitors' attention to the current health of the river, acting as a symbolic reminder of the entities—seen and unseen—that occupy these sacred spaces that demand our respect and care.

Bird decoys, once made of ceramic, were said to have originated as indigenous hunting aids and were used along the river Nile in Egypt almost 2,500 years ago. There are also accounts of decoys made of rushes, grass, and carved wood, used by Native Indian American communities in North America. Duck skins and feathers from fresh kills were sometimes stretched over the frame of the decoys, transforming them into even more lifelike representatives of birds.

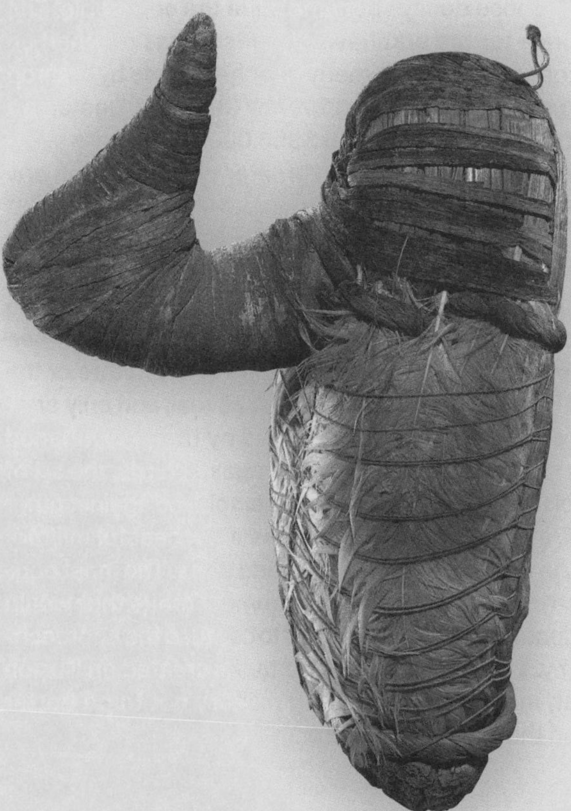

Duck decoy, ca. 400 BC–AD 100, Lovelock Cave, Nevada, photo: Smithsonian National Museum of the American Indian.

Early American colonists and settlers recognized the value of decoys and began using them for hunting and sporting activities. By the 1800s, professional decoy carvers began to emerge, creating highly detailed and realistic decoys using a variety of materials including wood, cork, and paper-mache.

Duck and goose decoys became quite the collectible. The record price for the purchase of a decoy was reached on the sale of a red-breasted merganser hen. Made by Lothrop Holme, the object was sold in 2007 by Guyette & Deeter in a joint auction with Christie's for $856,000. Decoys are available to collectors probably in greater numbers than any other form of folk sculpture.

Hunting decoys functioned to create a false sense of security for the real birds flying overhead, luring them to an area where they might not otherwise land in fear of danger. Deception and trickery are embedded in the nature of the object. The birds perceive the object as real, and only after landing close by are shot or trapped by the camouflaged humans lying in wait. Not only used against birds, the art of deception is a tactic also frequently employed against other humans, and was regularly used as a successful military strategy in the First and Second World Wars. The most popular decoys then were inflatable dummy tanks that would be positioned along borders, used to confuse the approaching enemy. The assembly of false tanks created an illusion of power, making the deceiving military out to be better equipped than they really were.

Decoying Wild Ducks—Lying low—A flock coming.
Public domain image acquired from
Library of Congress, Washington, D.C.

An inflatable dummy tank was used in a deception plan to mislead the Germans about Allied intentions in the run-up to the Normandy landings. Photo: Historycollection.com via Alchetron.

Maryland Canvasback Duck Decoy, early 20th century (artist unknown), Smithsonian American Art Museum, Gift of Herbert Waide Hemphill, Jr.

This art of deception—the act of luring one towards a trap, or otherwise off their intended course—was deployed in a groundbreaking conservation project entitled "Project Puffin," launched in the 1970s by biologist Stephen Kress on Eastern Egg Rock in Maine. By 1885, all the Atlantic puffins that had once been abundant on the island had disappeared as a result of excessive hunting by humans. Kress arranged for a small community of hand-painted wooden puffin decoys to be positioned on the island, and set up sound equipment that mimicked the social calls of the birds. Attracted by the decoys, puffins in flight that were heading past the desolate island began to land. On arrival, it did not take long for the newcomers to realize that their stiff-looking doppelgängers were fake, but it didn't seem to matter; the decoys functioned as a lure to show the birds that the island was safe, and in fact an ideal breeding ground. In addition, Kress made holes and tunnels for the birds to nest in, and after some years, a strong and healthy colony was established, which kept the birds returning year upon year. This system has now been replicated many times around the world as a successful way of re-establishing lost or diminishing bird colonies.

However, not all birds fall for the realism that humans attempt to construct through these winged sculptures. On Mana Island in New Zealand, a gannet named Nigel was dubbed "the loneliest bird in the world," as he became the only real bird living among a colony of concrete gannet replicas. The previous seabird colony had disappeared from the island forty years before Nigel arrived solo, seemingly the only one of his kind who was duped by the trap of the "healthy" decoy colony. Nigel lived alongside the concrete replicas for years, and was even documented by researchers stroking and making nests beside the sculpted birds, but naturally there was no reproductive activity, no young to be hatched. We can only guess whether Nigel was truly tricked or was just happy enough playing along.

***

Before darkness fell in Askeaton, as the rest of the group took to the local pub post-tour to celebrate the artworks, I crept back to the edge of the Deel to take one last look at the decoys. This is when I noticed the tall grey heron perching on one foot on a rock close to the decoys, perhaps curious to see what these odd-looking birds were up to. The decoys were bobbing awkwardly from side to side, moving more like the arm of a lucky Japanese cat in the rear window of a car than moving gracefully with the tidal current like I had hoped for. In Irish mythology, the image of a heron or crane standing on one foot signals their ability to have one foot in "our" world, and the other in the "under-" or "otherworld" — a parallel dimension that co-exists with that of our reality. These birds are perceived as bridging both worlds at their will, and as messengers of transcendence. It was often said that other entities from this otherworld would shapeshift into "our" world in the form of hares, goats, ravens, or eagles, and would bear messages, foretell events, bring omens, and trick and deceive us, all in attempts to warn and deter us from any disrespectful behavior we impose on the sacred land and resources we depend upon.

Suddenly the heron's other foot dropped down from beneath the tufts of its grey feathers, and it summoned the power in its large wings to lift itself off the small rock. It headed downriver, and I headed in the opposite direction towards the pub, to catch the others before last orders were called.

ENTRAPPING THE EYE
BY SUZANNE WALSH

When Bryony talks about her residency by the River Deel, I'm reminded of the river I grew up beside, the Slaney in Co. Wexford. Like in Askeaton, there are ruins of a Norman castle nearby, and similar wildlife too; herons, kingfishers, and other creatures frequent the waters. The River Slaney is also tidal, but instead of being named after a worm or a beetle like the Deel, it's said to be named after a Gaelic Chieftain, Slainghe, although some sources say it gets its name from the Irish for health, *sláinte*. As it progresses towards the sea, it flows out in wide bends that accommodate both marine and river life.

While much of the Slaney is now an area of special protection, there is a restricted duck hunting season. When reading about Bryony's decoy sculptures, I started thinking about my own experiences encountering decoys used by duck hunters on the river here. The hunters are neighbors of mine, and as I live right on the shoreline, I've been familiar with the occasional boom of guns upriver since childhood. Usually men, they come to shoot on Sundays between September and February, the day and months they are licensed to do so. They come either early in the morning or in the evening, in cars towing a trailer with dogs for waterfowl retrieval.

The hunters wear protective clothes printed in the latest camouflage patterns, which have progressed beyond the traditional black and dark green of the past. There's now a bewildering variety of styles on hunting apparel websites, designed to blend in with all kinds of terrain. In this case, it's the wilted greens and browns of autumn and winter. The hunters evade attention from ducks by disappearing into the background; they are merely a part of the vegetation. Their clothing disguises them so they can act through subterfuge; they might mimic the river bank, the dark dripping trees, the reeds amongst which they hide. The hunters also bring their plastic decoy ducks, but unlike the sculpted decoys that Bryony makes, these are mass-produced in a factory far away, and bought in hunting stores.

The hunters on the Slaney function as a community of their own—one that I'm outside of, although I live there—and I adjust my habits when they are present. I don't go out in my boat when they are shooting, but I often speak to them on the shore. They are friends with my elderly cousin, who is a former fisherman. When you live in a small community, you find ways to co-exist, and we talk about the tides, the particularity of winds on parts of the river, the forecast. But the hunters have strategies that they share only with each other, and a map of the river that is somewhat different to mine, with favored hunting spots and precise knowledge of both duck and river activities.

While not perfect, the Slaney is in good health, and is currently rich with many species, although there are some issues with too much fertilizer entering the water and causing the reedbeds to expand, changing the shape of the river. Seals swim up occasionally as they chase salmon, and ospreys have recently been sighted too. At night, the cry of the curlew echoes through the valley. The Deel, however, is less healthy, and Bryony's decoys highlight the pressing issue of pollution. She places her sculptures in order to be seen by the local Askeaton community (as well as visitors), so in a way, her viewers are her prey, as she draws their eye to the health of the water.

Her audience are perhaps initially attracted by the bird-like shapes, but the decoys quickly function as beacons of warning.

The duck hunters' plastic decoys on the Slaney highlight a different aspect of the river's health, signaling a body of water that is healthy enough to sustain a breeding population of ducks to hunt. However, this is artificially maintained, because the hunters feed the ducks regularly with oats and barley on the small platforms they build amongst the reeds. They slowly build up a sense of safety and routine for the ducks over time, so that they are unaware that eventually this sanctuary will become a site of danger for them. The decoys, along with the food, come as a kind of Trojan horse, a reassurance of safety that is actually a forewarning of death. What the ducks make of the decoys is hard to know. Hunters debate their strategies on online hunting forums. In one post I read advice that you should "never bunch up your decoys," as ducks bunch together when they are nervous. I read many threads on these forums fretting about ducks ignoring decoys or landing between them by accident. There is a lot of knowledge needed to set your ducks in a row, so to speak.

Just like the curation of an exhibition, the hunters must carefully arrange their decoys, their hides, their platforms. Hunters on one website discuss various possible formations that the decoys might be arranged in, for example "The Big River Combination Spread," "The Hidden Waterfowl Oasis," and "The Interspersed Goose Spread." This may not be art, but there is a craft to fooling birds.

When Bryony mentions the entities and animals of Irish mythology that shapeshift between worlds, it reminds me that, on the water, daylight can play tricks on the eye. Observing things and creatures at different times of the day, it often feels like other energies are at play. When I'm in my boat, things I see in the water can change shape, water being full of light and currents that cause this optical deception. My first encounter with duck decoys on the open water went like this:

I was in my boat, a traditional Slaney rowing cot, with my small terrier. We saw some ducks, nothing unusual about that, but this time their behavior seemed strange. They usually wait until you are close before flying up in a chorus of quacks, but these ones just bobbed along obediently in a line that moved from a reed bed outwards onto the river. We approached, carefully, my dog alert at the prow, but the ducks held their position. The more I looked at them, the more I felt that their movement seemed unnatural, and I was worried they were unwell. The ducks ignored my boat, and kept on moving out onto pale grey waves. Something felt uncanny, the gleam on their feathers was too dull, but then a shift in the light highlighted their carefully molded bodies. I realized we had been fooled by decoys, which had been arranged carefully so that they would float on the surface in a lifelike manner. In the meantime, the real ducks had flown to the other side of the river.

In Irish folklore, one must take care not to be tricked by the *Sidhe*, or the fairy people. They can pretend to be people you know in a bid to lure you away to the otherworld. Things are never as they seem, and when encountering their reality, land that is familiar can suddenly become unfamiliar; you might be in danger of getting lost. But the *Sidhe* also bring a richness to the landscape, a mystery and a shared narrative to live alongside. Which brings me back to Bryony's work in Askeaton, and the task that artworks like her sculptures undertake. Her decoy birds also cause a rupture in reality so that we can see something that we haven't before. The birds' artificiality becomes uncanny, and draws attention to that which can be hidden—in this case the pollution that an otherwise picturesque river contains. This trickery elicits uncertainty as well as awe, ultimately widening our perspective.

For a moment, Bryony's sculptures appear to be alive as real birds bobbing on the water, but quickly become something else, a sentry, or a beacon, a guardian of sorts. In the case of environmentally engaged projects like this one, an artwork can be a bridge between the abstract language of science and a local community and its visitors, bringing a more human or creative approach to the conversation. In Askeaton, this approach is crucial to highlight the very real issue of the pollution of waterways, and what that means for both people and wildlife. Bryony's decoys placed in the public, outside of the gallery, also bridge the gap between an often urban art world and a small rural community. The decoys lure us in to ask us to reposition ourselves, to see through the theater of the everyday. They ask how this theater can be remade, and how we can be part of that recreation.

DUCKS INTO HOUSES: DOMESTICATION AND ITS MARGINS

BY MARIANNE ELISABETH LIEN

The first thing you might notice as you arrive by boat to one of the many islands in the Vega Archipelago is a wooden cross carrying an inscription "FREDLYST". *Fredlyst* is an archaic proclamation that literally translates as peace. Sticking a wooden cross into the thin layer of soil on these remote Norwegian islands was an ancient way of announcing the presence of nesting eider ducks.[1] The cross signaled that somebody cared for the birds and that other people ought to stay away, or, if they needed to go onshore, that they must be very careful and quiet. The ancient crosses are long gone, but as the eider duck practice has been revived in recent years, crosses have emerged too.

Eider ducks have been cared for on the islands at least since Viking times, possibly much longer, and the seabirds have been hunted and harvested for as long as the coastline has been inhabited (Berglund 2009). During the past millennium, a specialized production has evolved in which locals build nests and "houses" for the birds and collect eiderdown and eggs in return. The seasonal

1 The Vega Archipelago is a cluster of approximately 6,500 islands off the coast of Nordland County, just south of the Arctic Circle. Since 2004, it has been listed as a UNESCO World Heritage site. http://whc.unesco.org/en/list/1143.

practice of attending to the rookeries and the birds constitutes one of the many landscape practices that were required to sustain oneself and one's family on these islands. In what follows, I draw on Norwegian anthropologist Bente Sundsvold's ethnography (herself native to Vega) and her detailed account of a peculiar practice that state legislation and ignorance nearly made extinct but that has recently been reestablished, due in part to international recognition associated with the islands designation as a unesco World Heritage site in 2004.

The annual encounters between ducks and their people take place in spring, when the eider (*Somateria mollissima*), locally referred to as *éa*, come onshore to nest. With thousands of islands to choose from (most of them uninhabited by people), they could go anywhere, but over the course of many generations, nesting ducks have found their way to where people are. Hence, during the brooding season, which lasts about a month, "people live in the midst of a colony of nesting eiders" (Sundsvold 2010, 2015). Carefully prepared, with roofs on top and dried seaweed inside, the nest boxes offer a site of protection against rough weather as well as predators.

Some nesting sites are placed against the wall of a house or within a cluster of buildings. Others are assembled on otherwise uninhabited islands and islets. Sundsvold's collaborator, Aud Halmøy, one of the pioneers of revitalizing eider duck practice, prepared more than 150 nests in one season, about three times as many, she reckons, as the number of nests that will actually be used. This way, the ducks will get to choose the nests they find most appropriate. Assembled from driftwood, leftover planks, slabs of stone, or simply an old boat turned upside down, such makeshift "domuses" are visual reminders of human presence in this remote island landscape.

Locally, as Sundsvold vividly describes, the brooding season had a special atmosphere and a special name: *varntie*. The prefix *varn* may be translated to "care," "precaution," or "protection," while *tie* refers to time (Sundsvold 2010, 96), but it is a word that few Norwegian speakers today would understand. As a "linguistic ghost" from another era (Mathews 2017), it enacts a landscape in which relations were done differently, when human activities were restricted by the presence of birds. *Varntie* was the time when everyone kept quiet, when dogs and cats were on leash or inside and when walking was restricted to the old established footpaths to which the birds were accustomed (Sundsvold 2010). Smoke from fireplaces was avoided, if possible (Fageraas 2016). Children were told to move slowly, speak softly, and pay attention. If they encountered a bird's nest, they were to gently step aside. Today, similar precautions are practiced by community residents. If you move carefully, or "think like a bird," as Aud Halmøy put it, they will slowly come to accept your presence. Others recall that those who cared for birds did not wash their clothes, in order to make it easier for the birds to become familiar with their people (Fageraas 2016). Sundsvold describes the process of caring for birds as a gradual earning of the birds' trust. The clue, she suggests, is to make oneself predictable to the birds. Achieving that requires an effort from everyone present, and the term *varntie* is a way of capturing this exceptional time. This is a term that cuts across distinctions between humans and their natural surroundings and instead fixes a shared temporal slot that is set apart for humans and birds alike.

But what attracts the ducks? For a human observer, the most obvious attractions are the nesting sites themselves. With a solid roof and dry seaweed bedding, they offer a ready-made comfort that nesting ducks can obviously appreciate.

Another key amenity is the protection against predators that this arrangement offers. The roofed houses make the nesting sites less visible to otters and eagles. But the rookery is only a partial protection, delaying the attack or diverting the predators' attention to nesting sites in the open. It offers no guarantee. This is where human presence becomes significant. Their mere presence can scare the predators away, and active human intervention can make a difference, too: now and then, the silence of *varntie* is shattered by a gunshot in the air. Aud Halmøy always brings a gun with her on the daily rounds. Occasionally a crow or a black-backed gull is shot. And before the season starts, mink traps are distributed to avoid the devastating effects of a recent predator.

Are the eiders tame? Bente Sundsvold insists that they are not. This, she argues, is why the *varntie* was so important. The idea is that a successful brood of ducklings will cause the female eider to return year after year. Later, as they mature, the ducklings will return, too, securing additional income. When Aud Halmøy lost nearly the entire stock of eiders due to a particularly nasty predator attack at the rookeries she had prepared on an uninhabited islet, she was devastated. Certain that years of effort to earn the trust of the ducks was lost, she was almost ready to give up. But much to her surprise, the eider ducks returned in large numbers the following year, though their preferences had changed: instead of nesting on the slightly remote islets, they now chose to nest near the cluster of houses. They chose the rookeries that would be much less quiet but that would offer more protection against predators, due to the continuous presence of Aud and her family. Perhaps, then, the predators too are a necessary part of this domestication assemblage? Sundsvold claims that they are: without predators, she argues, it is unlikely that the birds would have sought these rookeries in the first place. In this way, the presence of otters and eagles facilitates the human gathering of eiderdown and eggs. Another key component is some form of memory across bird lives and generations: As is the case for salmon (Ween and Swanson 2018), the relation of domestication hinges on migratory returns of the same birds, year after year, and then, later, of their offspring. To build up a *dunvær* takes many, many years; it is a gradual institutionalization of trust, articulated through the brooding *éa*'s selection of house, and passed down through generations of birds, as well as people.[2]

DOMESTICATION RECONSIDERED

Seen through a conventional approach to domestication, these eider duck arrangements are barely visible. Like many other instances of multispecies relations in the North, the story of the eider ducks challenges our categories of wild and domestic. Neither confined nor controlled—neither owned nor fed—the ducks have none of the characteristics that place them in the category of conventional domesticates (cf. Clutton-Brock 1994). And yet, the assemblages have been consequential for birds as well as for people. According to biological research on eider ducks on Spitsbergen, it is quite likely that human-made protected shelters have had an impact on the duck populations' survival rates. If so, it might also historically have changed selection pressures, possibly favoring those birds that tolerated human presence. For humans, it has

2 *Dunvær* is the local name for the duck-rookery-down assemblage. *Dun* means down (feather), and *vær* is a common name for sites associated with particular affordances, such as *fiskevær* (where fish is landed).

provided food as well as income and thus contributed to a more robust livelihood in a coastal landscape where fish were abundant but farming was precarious, pasture was scarce, and grain cultivation nonexistent.

Seen in relation to more recent debates on domestication, however, the duck and down assemblages at the Vega islands are a highly relevant case to consider, and one that can teach us about a kind of variation in domestication practices that is often overlooked. We see the unfolding of a social relation that is both asymmetrical *and* simultaneously built on trust. We see how presumably wild birds seek human compounds for protection, like so many animals have done before them. Finally, we see how the relations that unfold involve a complex interspecies relation of ducks-predators-humans, a relation that challenges the assumption of domestication as a binary process. Perhaps the relation is best described as what Natasha Fijn, with reference to Mongolian herding practices, calls a codomestic relationship: "the social adaptation of animals in association with human beings by the means of *mutual* cross-species interaction and social engagement" (Fijn 2011, 19, emphasis in original). According to Fijn, relations of codomestication are spatially situated, encompassed within the landscape, or codomestic sphere, which in the case of the Mongolian herders is synonymous with their encampment (Fijn 2011, 220).

According to historical sources, humans have gathered eiderdown in Vega for several hundred years, possibly for more than a millennium. Does this qualify as an incipient phase in a "gradual process of domestication"? Is it a phase on a journey from wild to domestic, from trust to domination, or from being "free as a bird" to being subject to tightened regimes of control and confinement? Based on Sundsvold's and others' accounts, we may reasonably conclude that it is not. That "ducks go into houses" does not mean that they have come to stay, nor that they are, in evolutionary terms, on a trajectory toward a final destination (cf. Zeder 2012, 91). Their makeshift domus is a temporary amenity, a material expression of a fragile relation of codomestication involving birds and people that offers no guarantees for anyone and that can easily fall apart. For Mongolian herders, these relations are spatially situated within their encampment. In Vega, the relation is also encompassed temporarily, in the specific time known as *varntie*. In this way, the *dunvær* assemblages remind us that domestication is—perhaps always—a precarious achievement, a sensibility that is easily lost when our assumptions about domestication are too committed to notions of confinement and control. They show that we should not let hegemonic idioms of European farming narrow our gaze.

This is important everywhere but especially at northern sites like Vega, where conventional farming is hardly enough to feed a family. In such regions, livelihoods have typically relied on a broad range of landscape practices and on subtle of ways knowing the land and the animals in their surroundings.[3] That some such practices were institutionalized is therefore not surprising. As Sundsvold explains, *fredlysning* means to publicly announce the consecration of peace, "in the medieval sense of a public proclamation" (Sundsvold 2010).[4] She describes how *fredlysning* sites have been registered in the Vega islands since the eighteenth century, but that the practice was probably much more common than these

3 This analysis focuses on coastal Northern Norway, but similar arguments could be made for other subarctic coastal regions, including Iceland, Scotland, the Faroes, and parts of Russia.

4 In old Norse language, *frid (fred)* connotes not only the absence of strife but also a relation of honor among free men.

legal registrations indicate.[5] *Fredlysning* was not a proclamation of ownership in the sense of announcing private property but rather a temporary institutionalization of specific user rights and user interests that regulated the relations between people and environmental resources in a region where farming and private property were of limited importance. The institution was still practiced in Vega in the 1950s and 1960s. Most of the proclamations of *fredlysning* involve eider ducks, so-called *dunvær*, but some concern cloudberries and other amenities, such as specific fish used for bait, grass that can be cut and used as fodder, and in the old days even otter, which was hunted for fur. As *fredlysning* did not require ownership of land, it served to secure a certain income or access to resources for people who were otherwise disenfranchised or had limited sources of livelihood.

During the twentieth century, the Norwegian state had made its presence more strongly felt. Especially from the 1950s onward, state efforts at centralization were commonplace and rather efficient. Anthropologist Ottar Brox voiced a solid critique of state policy in Northern Norway during the postwar era. His key argument is that through various subsidy regimes that sought to enhance capital growth by favoring those who were specialized as farmers only, or fishermen only, the state ignored the uniquely adaptive subsistence strategies of fishermen-farmers of the Northern Norwegian coast (Brox 1966, 66). This made traditional livelihoods increasingly difficult to sustain. The push toward urbanization was further exacerbated by the state's classification of coastal hamlets as either places one should move to (*tilflyttingssted*) or places one

5 *Fredlysning* is mentioned several places in legal documents from the Viking era, such as in the *Frostatingsloven*, which is known to exist at least since the year 1000 and which was written down in 1200.

ought to move away from (*fraflyttingsted*), accompanied by state subsidies for those who decided to move (Sundsvold 2010).[6] The largest village on the central island in the Vega Archipelago was a place to move *to*, while all other settlements on the remote islands were places to move *from*. In the same period, electrification, concentration of schools, and the acquisitions of motorized fishing vessels all contributed to making the more remote islands less attractive. In spite of such measures, many remained, and the main exodus from the islands to the new center did not happen until the 1970s—not until the arrival of mink farming.

By the 1970s, state-subsidized mink farming had been promoted by regional developers, and a number of farms had been established, even in the Vega islands. Mink (*mustela vison*) were introduced to Norway from North America in the 1930s, and the first farms were established in the coastal region of Helgeland in the 1940s. Soon, a feral population of mink was established, too, and according to Sundsvold, its impact on the domesticated ducks was catastrophic, as feral minks became a new and efficient predator. Why, she asks, did the islanders not resist the establishment of mink farms side by side with eider duck houses? Why was it so easy to place wild animals in cages?

The answer, she suggests, has to do with notions of progress and the calls for rural economic development that were the mantra of the postwar era. Mink farming was potentially profitable, mobile, and scalable and could provide remote communities with the "development" that they sorely needed. I would add that the fact that mink

6 The subsidies began in 1950 and remained for the next thirty-seven years, dismantled just fifteen years before the application to grant the Vega islands World Heritage site status (Sundsvold 2015).

were confined, controlled, and provided fur that could fetch a good price on the global market made it an obvious commodity for the imagined postwar future. That it simultaneously undermined another ancient livelihood, another precarious form of human-animal domestication, was not enough to keep it out.

Did the state completely ignore the presence of the ducks on the islands and their importance for the islanders? A final historical shift indicates that this might have been the case. In 1981, after having been inscribed in legal regulations for 140 years (and informally practiced much longer), the ancient institution of *fredlysning* was suddenly abolished. The occasion was a new set of regulations for hunting game.[7] If the introduction of the mink was catastrophic for the traditional duck and down rookery assemblage, this was, according to Sundsvold, the final nail in the coffin. With the implementation of the new regulations, the few remaining nesting sites that were still maintained could no longer protect the birds during brooding season. From now on, the landscape was open for all citizens on equal terms, as state-owned territory tends to be in Norway, where the right to roam is an important principle securing access.[8] The change signals a recognition of Norwegian landscapes as fundamentally state-owned and equally accessible to all, as well as an emergent recognition of landscapes as recreational wilderness, known in Norway simply as *natur* (nature).[9] As *natur*, the Vega islands and all its inhabit-

7 "*Lov om jakt og fangst av vilt, 28. mai 1981.*"

8 This is anchored in the legal term *allemannsrett* (all-man's-right).

9 This right-to-roam had already been legally established in 1958 with the so-called *Friluftsloven* (outdoor life legislation). See https://lovdata.no/dokument/NL/lov/1957-06-28-16. The new hunting regulations confirmed the principle of general access for hunting and fishing, as well.

ants are no longer exclusively known and cared for by locals but by the state Ministry of Environment, protected through policies of nature conservation and known through natural science. This is in contrast to agricultural land, husbandry animals, and pasture, which are regulated by the Ministry of Agriculture. In this way, human-animal relations that were not recognized as husbandry, but that were also not "not-husbandry," are matter-out-of-place, ignored or marginalized. Similarly, locally recognized connections between certain kin groups and certain cloudberry marshes that were once proclaimed through *fredlysning*—another subarctic practice of food procurement that does not fit the model of modern agriculture—are also ignored. In other words, as the Vega islands have become part of a state-managed Nature, they have simultaneously been alienated from the practices that once shaped people's livelihoods, practices that do not operate according to the binary of Nature and Culture.[10] At odds are two landscapes and two sets of human-animal relations: one is fluid, precarious, fragile, and largely invisible for state developers; the other is defined, protected, highly visible, and classified according to the hegemonic binary of the wild (which can be harvested, hunted, and used as a site for recreational pleasure) and the domestic (which belongs in cages). The brooding ducks' attentive clumsy steps toward makeshift houses are matter-out-of-place, prey to the feral mink, and invisible to the new hunting regimes. The fragile relations

---

10 This situation bears resemblances to the cultivated forests in Northern Italy. According to Andrew Mathews (2017), G146: "Over the last one hundred and fifty years, industrialization, rural out-migration, the arrival of alternative forms of fertilizer, and the arrival of successive epidemic diseases have undermined chestnut cultivation and transhumance, leaving a ruined landscape that is haunted by material and linguistic ghosts."

of codomestication that made the islands a codomestic sphere are made absent through the state regulation of outdoor life through equal right-to-roam for all.

REFERENCES

Berglund, Birgitta. 2009. "Fugela Federum in Archeological Perspective: Eider Down as a Trade Commodity in Prehistoric Northern Europe." *Acta Borealia* 26, no. 2: 119–35.

Brox, Ottar. 1966. *Hva skjer i Nord-Norge? En studie i norsk utkantpolitikk. (What Happens in Northern Norway? A Study of Policy for the Periphery)* Oslo: Pax.

Clutton-Brock, Juliet. 1994. "The Unnatural World: Behavioral Aspects of Humans and Animals in the Process of Domestication." In *Animals and Human Society*, ed. A. Manning and J. A. Serpell, 23–36. London: Routledge.

Fageraas, Knut. 2016. "Housing Eiders, Making Heritage: The Changing Context of the Human-Eider Relationship in the Vega Archipelago, Norway." In *Animal Housing*, ed. K. Bjørkdahl and T. Druglitrø, 82–99. London: Routledge.

Fijn, Natasha. 2011. *Living with Herds: Human-Animal Coexistence in Mongolia*. Cambridge: Cambridge University Press.

Lien, Ween, and Swanson, 2018. *Domestication Gone Wild: Politics and Practices of Multispecies Relations*. Durham: Duke University Press.

Mathews, Andrew. 2017. "Ghostly Forms and Forest Histories." In *Arts of Living on a Damaged Planet*, ed. Anna Tsing, Heather Anne Swanson, Elaine Gan, and Nils Bubandt. Minneapolis: University of Minnesota Press.

Sundsvold, Bente. 2010. "Stedets herligheter — Amenities of Place: Eider Down Harvesting through Changing Times." *Acta Borealia* 27, no. 1: 91–115.

Sundsvold, Bente. 2015. "Den Nordlandske Fuglepleie': Herligheter, utvær og celeber verdensarv." PhD Dissertation, University of Tromsø.

Zeder, Melinda. 2012. "Pathways to Animal Domestication." In *Biodiversity in Agriculture: Domestication, Evolution, and Sustainability*, ed. P. Gepts, T. R. Famula, R. L. Bettinger, S. B. Brush, and A. B. Damania, 227–59. Cambridge: Cambridge University Press.

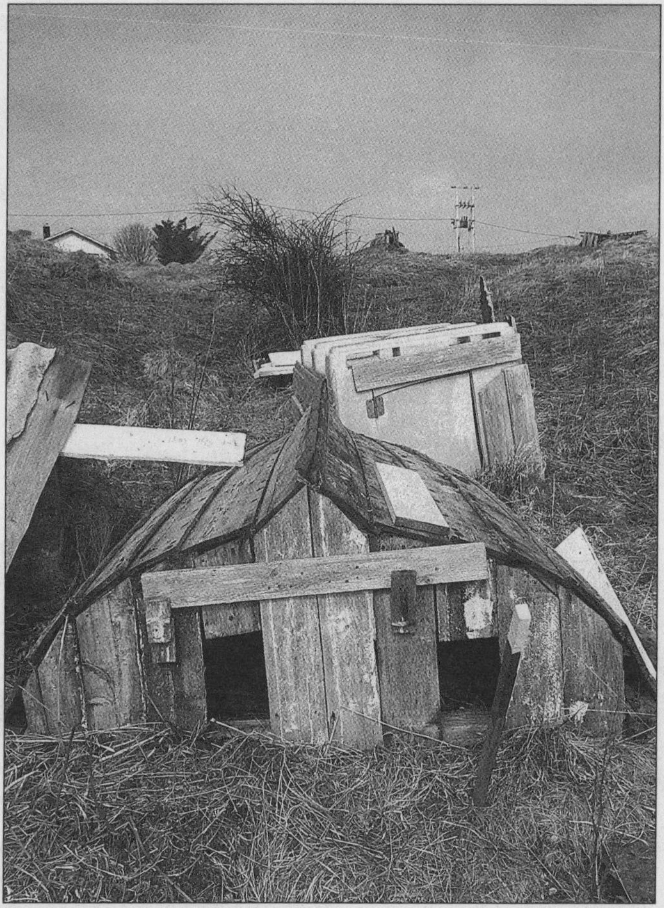

Eider duck houses. Photo: Eva Bakkeslett.

Coda:

Echoes and Omens

FALLEN ANGELS:
Birds of Paradise in Early Modern Europe

By Natalie Lawrence

When birds of paradise first arrived in Europe, as dried specimens with legs and wings removed, they were seen in almost mythical terms—as angelic beings forever airborne, nourished by dew and the "nectar" of sunlight. Natalie Lawrence looks at how European naturalists of the 16th and 17th centuries attempted to make sense of these entirely novel and exotic creatures from the East.

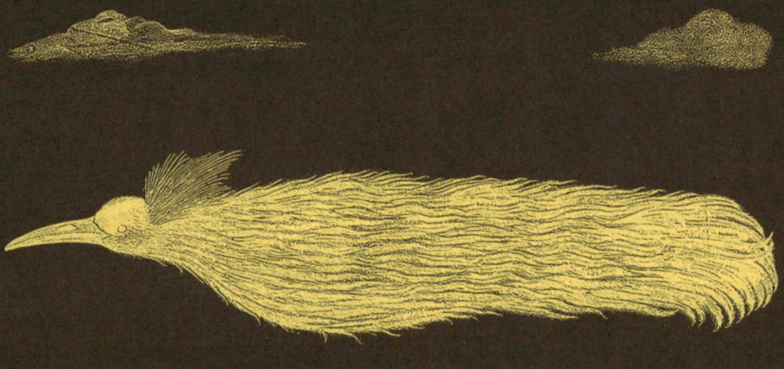

Bird of paradise from Ulisse Aldrovandi's *Ornithologia* (1599), via Wikimedia Commons. Public domain.

In 1522, the dismembered skins of new and fabulous creatures were brought to the court of Emperor Charles V of Spain. These were birds of paradise from the East Indies, carried with a cargo of spices and other valuable marvels on the last remaining ship from a fleet that had left in 1519 to circumnavigate the globe under Ferdinand Magellan. These birds were unknown and entirely unseen in Europe, but they caused an instant stir. The ship's chronicler, Antonio Pigafetta, described how five of these enigmatic objects were presented to the Spanish sailors by a Moluccan Sultan as a royal gift for the Emperor:

> *These birds are as large as thrushes; they have small heads, long beaks, legs slender like a writing pen, and a span in length; they have no wings, but instead of them long feathers of different colours, like plumes . . . they never fly, except when the wind blows. They told us that these birds come from the terrestrial Paradise, and they call them 'bolon dinata', that is, "divine birds".*[1]

These unusual skins were shrunken and wingless, causing their beaks and gorgeous plumes to be disproportionately exaggerated. They were prepared by New Guinean hunters for tribal dances and displays, which still occur in tribal areas of New Guinea. The flesh and limbs were removed, though methods varied by tribe and species of bird, and the whole skin was smoked to heighten the dramatic effect of the elaborate feathers.

1 Pigafetta, Antonio. *The First Voyage Round the World by Magellan, Translated from the Accounts of Pigafetta and other Contemporary Writers (Works Issued by the Hakluyt Society).* H.E.J. Stanley (ed.). (New York. Cambridge University Press, 2010), 143.

Yet the living birds were virtually unknown to anyone outside New Guinea, even to the Moluccan traders who dealt in New Guinean products. One Portuguese apothecary who travelled around Southeast Asia wrote in 1513 that birds "which are prized more than any others come from the islands called Aru (Daru), birds which they bring over dead, called birds of paradise (*pasaros de Deus*)". Bird of paradise plumes, in particular, were some of the most coveted products in Asia and had been part of Asian trade networks, which circulated spices and other valuable goods such as ivory and textiles, for at least 5000 years before Europeans reached the region in the late fifteenth century.

Thought to be the first colour depiction of a bird of paradise, by Croatian miniaturist Julije Klovicín from the exquisite *Farnese Hours*, 1546, Morgan Library and Museum via Wikimedia Commons. Public domain.

Spices such as cinnamon, cloves, and nutmeg grew only in the Moluccas, but reached the Middle East and eventually Europe in the Middle Ages through trade networks that passed through Venice. Before the sixteenth century, however, nothing was known of the birds of paradise in Europe. It was this high demand for spices in early modern Europe that drove the first European explorations of "a strange, and for so many ages, an unknown world, in order to search for the islands where spices grow", and, more importantly, to profitably undercut the Venetian monopoly.

Spanish court secretary Maximilianus Transylvanus wrote an account of Magellan's travels, *De Moluccis Insulis* (1523), constructed from interviews with sailors. It included Islamic Asian myths transported to Europe along with the skins:

> *The kings of Marmin began to believe that souls were immortal a few years ago, induced by no other argument than that they saw that a certain most beautiful small bird never rested upon the ground nor upon anything that grew upon it; but they sometimes saw it fall dead upon the ground from the sky. And as the Mahometans, who travelled to those parts for commercial purposes, told them that this bird was born in Paradise, and that Paradise was the abode of the souls of those who had died, these kings embraced the sect of Mahomet, because it promised wonderful things concerning the abode of souls. But they call the bird Mamuco Diata [Bird of God]...* [2]

Accompanied by such fabulous accounts, these striking objects could hardly fail to become something special. In the 1540s, more bird of paradise skins began to trickle

2 Transylvanus, Maximilian. 'A Letter from Maximilianus Transylvanus to the Most Reverend Cardinal of Salzburg...'. In Stanley, H.E.J. (ed.), *The First Voyage Round the World by Magellan, Translated from the Accounts of Pigafetta and other Contemporary Writers, Works Issued by the Hakluyt Society*. (New York: Cambridge University Press, 2010), 205–6

into Europe from Portuguese and Spanish trading vessels returning from the East Indies. The skins, most of which bore the iridescent green and velvet brown plumage of the greater bird of paradise (*Paradisaea apoda*), were prepared for trade, limbs removed, and rapidly acquired by aristocratic collectors for their curiosity collections.

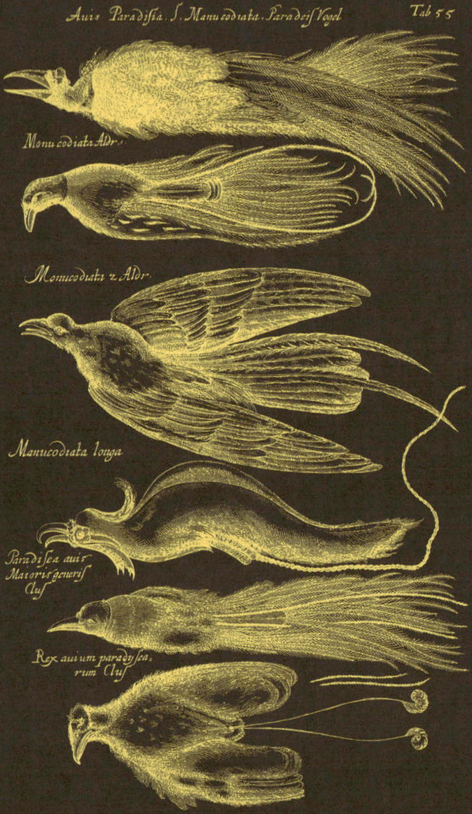

Illustration of various birds of paradise skins from Jan Jonston's *Historiae Naturalis De Avibus Libri VI* (1650) via Wikimedia Commons. Public domain.

"Chamaeleon Aereus", illustration from Francesco Calzolari's *Musaeum Calceolarianum Veronense* (1622). Public domain.

There were still a relatively small number of specimens scattered in a variety of collections across Europe. Apart from the collections of the Cardinal of Salzburg and Charles V in Spain, the German humanist Conrad Peutinger owned a specimen, as did the Italian scholar Julius Caesar Scaliger. The apothecary-collector Francesco Calzolari in Verona had a specimen of a *chamaeleon aereus* described in his collection catalogue, and the German physician-botanist Melchior Weiland had acquired a specimen during his travels in the Levant. The skins fetched painfully high prices on the open market: a 1550 pamphlet from Nuremberg advertised a skin of the birds known to the Greeks as "*apodes*" ("without feet"), on sale for the price of 800 thalers, equivalent to a year's salary for a university scholar or skilled musician.

Very little was published about these rare birds until the middle of the sixteenth century. Even so, it didn't take long for their depictions in books of emblems, natural philosophies, and in many other genres to transform them into angelic, floating creatures. These images were based in part on the Islamic myths published by Transylvanus and the fact that the "Orient" has been seen as an exotic paradise full of wonders and riches for hundreds of years. It was almost no surprise that the East should turn out to be a Paradisal land full of spices and jewel-like birds.

"Apis indica". The bird of paradise enshrined in a new southern constellation by Dutch astronomer Petrus Planchius, and here shown in a detail from Bayer's *Uranometria* (1661). Public domain.

Few authors actually saw the bird of paradise skins, so closely did collectors keep them from public view. And none, of course, saw the living "angels" they described. The ideas that the birds were like angels did not in fact originate with the legless nature of the skins. This leglessness was used, rather to support the fantastical life histories that authors imagined. The birds of paradise were literally mythologised.

The first pivotal account of the *manucodiata* was in the *Historiae Animalium* (1551–58) of Swiss naturalist and physician Conrad Gessner. The five volumes of this gargantuan work followed the Aristotelian structure of the natural world and is a seminal example of a Renaissance encyclopedia. It combined image with text for the first time, including about 1200 woodcuts. The complete "history" of each alphabetically-ordered creature was arranged in a set of lettered thematic sections such as synonyms, form, and uses. The size of each entry depended on the volume of extant material on it: newer creatures merited smaller entries than well-known ones.

In order to form a complete history of the birds, Gessner brought together as many sources as he could. He included one of the first extant scholarly descriptions of a bird of paradise, found in the expansive *De Subtilitate* (1550), a book by the Italian mathematician and astrologer Jerome Cardan. Cardan reasoned that, because these birds were never seen alive and could not land without feet, they must exist perpetually airborne in the highest reaches of the sky. Cardan argued that nothing solid was ever found in their bodies, so they must be like the mythical rhyntace, "a little Persian Bird which has no excrement, but is all full of fat inside, and the creature is thought to live upon air and dew". He also suggested that males had a cavity in their backs in which females laid eggs and incubated them, and he coined the Latinate term *Manucodiata*, drawn from the Malay name *Mamuco diuata* (birds of God). Gessner also used the description by French naturalist Pierre Belon, who, in his *L'Histoire de la nature des oyseaux* (1555), described the head-dresses of the Janissaries he had seen in the Levant. They contained "plumes of a bird called the *Rhintace* ..." from "a small creature of which only the skin is left" that he believed "may be the *Phoenix*".

On the basis of these and other accounts, Gessner developed some marvelous ideas about the life histories of the birds. He suggested that these "*Lufftvogels*" (birds of the air) were effortlessly suspended by their haloes of plumes. Gessner imagined several uses for these hair-like projections, or *cirri*—the naked shafts projecting beyond the rest of the feathers. He speculated that the birds might use them to hang from tree branches when they were tired. They might also be vital to the birds' love lives: Gessner envisaged couples entwining their cirri together when mating, and as the female sat atop the male to brood her eggs amongst the clouds. He maintained that the fatigue of unending flight might in fact be taken away by the birds' constant movement, like the perpetual movement of a clock pendulum. He argued, though, that they could not possibly live only on pure air or ether. It was far too thin to sustain anything. They were birds, after all.

Gessner often called upon correspondents to get information or images he needed. Conrad Peutinger in Augsberg sent Gessner a report and drawing of his own personal specimen to be the basis for the woodcut illustration featured in Gessner's book. The skin in the resulting woodcut is apparently legless but lacks the hair-like projections that figured so in Gessner's theories and were clear in other contemporary images such as Calzolari's.

612 De Auibus

¶ Sunt qui hanc auem Germanicè Lufftvogel appellent, hoc est auem aeris: siue quòd in aere semper ferè degat, siue quòd eo etiam uiuere uulgò existimetur. Quidam receptaculum putant sub alis habere fœminam, ubi oua foueat. ¶ Reges Marmin (in Moluccis insulis) paucis antè annis immortales animas esse credere cœpere: haud alio argumento ducti, quàm quòd auiculam quandam pulcherrimam, nunquam terræ aut cuiquā alij rei quæ in terra esset insidere animaduerterent, sed aliquando ex summo æthere exanimem in humum decidere. Et cum Mahumethani qui ad eos commercij causa commearent, hanc auiculam in paradiso ortam, paradisum uerò locum animarum quæ uita functę essent, attestarentur, induerunt hi reguli Mahumethi sectam, quòd hæc de hoc animarum loco mira polliceretur. Auiculam uerò mamuco diata, id est auiculā dei appellant: quam adeò sanctè religioseq; habent, ut se ea reges tutos in bello existimēt, etiamsi suo more in prima acie collocati

Bird of paradise from Conrad Gessner's *Historiae Animalium* (1551–58), from Biodiversity Heritage Library. Public domain.

Gessner's work was source material for many later publications, including Bolognese apothecary Ulisse Aldrovandi's *Ornithologiae hoc est de avibus historia* (1599).

Aldrovandi had several skins in his collection and differentiated four different species of *manucodiata* on this basis: *Manucodiata prima*, *secunda*, *cirrata*, and *hippomanucodiata*, each depicted in woodcuts drinking dew and floating amongst clouds.

Bird of paradise, shown here drinking rain, from Ulisse Aldrovandi's *Ornithologiae* (1599), from Biodiversity Heritage Library. Public domain.

However, in his text, Aldrovandi argued that the birds could not possibly live on dew alone and conjectured that their "sturdy beaks" were very like those of woodpeckers, and "very fit to strike insects". He also suggested that the cirri might aid "swifter flight" rather than being used in mating.

The birds certainly weren't confined to natural histories. They were used in many books of emblems such as Joachim Camerarius' *Symbolarum et emblematum* (1596), encapsulating ideas of spiritual ascension, lofty thinking, and restless, mercurial thought with their unending flight.

Felices nimium quorum ſuper æthera mentes,
Sublatæ cuncta hæc infera deſpiciunt.

Emblem featuring a bird of paradise from Joachim Camerarius' *Symbolarum et emblematum* (1596), from Internet Archive. Public domain.

The birds, with their elaborate plumes and flighty nature, were even used as emblems of the fickle and coquettish nature of vain and richly adorned women. They also made it into the stars: Dutch astronomer Petrus Planchius enshrined the bird of paradise in a new southern constellation, *Paradysvogel Apis Indica*, in his 1598 celestial globe.

In art, the birds' exotic and ascendant nature was powerfully symbolic. In Jacques Linard's painting *The Five Senses and the Four Elements* (1627), for example, a bird of paradise skin is depicted as if it is flying out of the window to escape the chaos of the study. It acts as part of an alchemical allegory, symbolising the pure aerial elements: these birds were too close to the heavens to be contaminated with the exigencies of the everyday world.

The Five Senses and the Four Elements (1627) by Jacques Linard via Wikimedia Commons. Public domain.

In Roelandt Savery's *Landscape with Birds* (1628), the birds are comet-like streaks in the sky, far above the terrestrial waterfowl and songbirds.

From the early seventeenth century some European naturalists got access to new kinds of skins that possessed legs. Several wrote long and disparaging refutations of earlier authors and their belief in legless wonders. The birds of paradise were becoming terrestrialised, just birds, no longer the ethereal angels they had been.

Detail from Roelandt Savery's *Landscape with Birds* (1628) via Wikimedia Commons. Public domain.

Yet the exoticism and moral connotations of the birds-as-angels meant that these images did not die out. Depictions of the birds were everywhere …—in paintings, religious texts, allegories, even in some newly-published retrograde natural histories. It was not only that the birds had become a geographical symbol of the wondrous East or a moral symbol of a pure existence, but that the image of monstrous legless birds fascinated people. This has not lessened today; they continue to enthrall, perhaps even more so as fabulous natural wonders than angelic impossibilities.

John Berger was an English writer, artist, and thinker whose work spanned fiction, criticism, and visual art. Grounded in Marxist humanism and a deep attentiveness to everyday life, his writing explored the politics of seeing, the dignity of labor, and the entanglements of people and place. From *Ways of Seeing* to *A Seventh Man*, he examined how images, histories, and economies shape perception. His novels, essays, and collaborations often centered the rural and the dispossessed, weaving tenderness with radical critique. Berger's work invites us to look more slowly and closely—at art, at the land, and at each other.

Ignace Cami is a Belgian interdisciplinary artist who breathes new life into old things. Originally trained as a printmaker and sculptor, he soon developed more active and social ways of disseminating his projects. Firmly rooted in folk culture, his work includes objects, sculptures, installations, and texts. He looks for things that, through the passage of time, have become disconnected from everyday life, and creates ways in which they can be reconnected in the present. Playfully experimenting with different forms of cultural heritage, Cami's practice displays a characteristic generosity, often inviting people to participate in his projects.

Sara Sejin Chang (Sara van der Heide) has been developing her multidisciplinary practice since the late 1990s, spanning film, text, immersive installations, performance, and painting. Her work combines spiritual evocation with historical research and the unraveling of colonial narratives, creating spaces for historical repair, healing, and belonging. Chang critically engages with Eurocentric systems of categorization and racialization, exposing their reach across contemporary Western life. Her poetic, intimate gestures center a meta-cosmic, inclusive approach to modernity, reimagining value and time. Through her practice, Chang opens up transformative perspectives on identity, history, and the sacred.

Monika Czyżyk is a Polish visual artist based in Helsinki, Finland, whose work moves between video, VR, painting, and performance to explore elemental relationships and collective processes. Working with what she calls "dirty media," she gathers materials—both digital and organic—to channel and transform energies across places, people, and technologies. Her practice is shaped by travel-led and site-specific research, anchored on Vartiosaari Island, which she considers a collaborator. Drawing on synchronicity and sensory experience, her installations and time-based works act as subtle bridges between cultures and beings, exploring how shared rituals and material gestures can foster connection across perceived divides.

Bryony Dunne is an Irish visual artist and filmmaker whose work blends documentary, fiction, and ecological inquiry to explore the entangled relationships between humans and the natural world. Through research-led narratives, she examines fantasies of control and the mechanisms of surveillance imposed on both people and ecosystems. Her films and installations often inhabit speculative or fact-based futures, using cinematic language to disrupt anthropocentric perspectives. Working across image and object, Dunne invites critical reflection on how borders—geopolitical, biological, or conceptual—are constructed and policed, while suggesting more fluid, interconnected ways of imagining our place within shared planetary systems.

Daniel Godínez Nivón is a Mexican visual artist, educator, and independent researcher based in Amsterdam. His practice merges social participation, education, and collective knowledge, employing methodologies like *tequio*—a traditional form of communal work rooted in Indigenous communities, especially in Oaxaca. Godínez Nivón explores the relationship between dreaming and nature, investigating how the dreams of humans and other beings can deepen our understanding of environmental issues and inspire collective imagination. Through projects like *Tequiografías*, he collaborates with Indigenous communities to create alternative educational materials that challenge hegemonic knowledge systems and promote situated Indigenous knowledge. His work invites reflection on the intersections of art, ecology, and communal learning.

Daisy Hildyard is an English writer whose work explores the entanglements between humans and the natural world, the porous boundaries of the body, and the traces of history in everyday life. Her fiction and essays weave personal observation with ecological thought, often reflecting on how global systems shape local experience. In her writing, the rural and the planetary coexist, animals and humans share space, and memory links past and present. Through both lyrical storytelling and quiet inquiry, Hildyard invites readers to reconsider the idea of separateness—between species, places, and selves—in a time of environmental and social urgency.

Manjot Kaur is an Indian visual artist whose paintings, drawings, and time-based media explore the sovereignty of ecology and women's bodies. Her practice weaves together speculative fiction, archetypal allegories, and precarious ecologies to challenge anthropocentric narratives. Drawing from ancient mythologies and histories, Kaur reflects on the relationships between humans and more-than-humans, envisioning multi-species futures. Her works delve into intimate worlds, encompassing the anthropology of wonder and awe, and respond to ecological grief through acts of care and kinship. By reimagining traditional symbols and narratives, she invites viewers to consider alternative ways of relating to the natural world.

Natalie Lawrence is a London-based writer, illustrator, and historian of science whose work explores the shifting boundary between nature and culture. Fascinated by the creatures we imagine as well as those we classify, she investigates how myths, fears, and desires shape human understandings of the natural world. Drawing on her background in zoology and the history of science, Lawrence traces how monsters, exotic animals, and plants have been seen as mirrors of human psychology and power. Her writing brings scientific insight into dialogue with storytelling, revealing the deep entanglements between ecology, imagination, and what it means to be human.

Ursula K. Le Guin was a visionary writer whose speculative fiction reimagined the boundaries of gender, power, and human connection. Influenced by Taoism, anthropology, and feminism, her novels—including *The Left Hand of Darkness* and the *Earthsea* series—explored alternative societies and ecological balance. Le Guin's work challenged genre conventions, emphasizing storytelling as a means to question dominant narratives and envision just, sustainable futures. She believed in the transformative power of imagination, asserting that "resistance and change often begin in art." Through her profound narratives, Le Guin invited readers to consider new possibilities for living and relating in the world.

Marianne Elisabeth Lien is Professor of Social Anthropology at the University of Oslo. Her research explores human–nature relations, domestication, food politics, and contested landscapes, with a regional focus on the Nordic Arctic and Tasmania. Lien investigates how life is sustained through materials—foods, tools, animals, and plants—amid environmental and colonial pressures. She has examined food systems from kitchen tables to fish farms, and is currently researching digitalization in reindeer husbandry. Her work advances collaborative, experimental ethnography and contributes to the fields of environmental anthropology, kinship, and food studies. Lien's scholarship critically engages with the Anthropocene, the commons, and decolonial futures.

Marjolein van der Loo is a Dutch curator, researcher, and artist whose practice explores ecological relationships, collective learning, and embodied knowledge. She develops long-term, collaborative projects engaging with plants, landscapes, language, and feminist and decolonial thought. As Curator of Contemporary Art and Heritage at Museum Het Nieuwe Domein and Associate Curator at Onomatopee, she weaves storytelling, sensory experience, and critical reflection into exhibitions, workshops, and publications. Her work nurtures alternative ways of relating—to each other, to place, and to the more-than-human world—centering care, reciprocity, and resistance to extractive systems.

Nicholas Mirzoeff is a visual culture theorist whose work interrogates how images shape power, identity, and resistance. As a founder of visual culture studies, he explores the politics of seeing in contexts ranging from colonial history to contemporary social movements. His writings, including *The Right to Look* and *White Sight*, examine how visuality enforces systems like white supremacy and settler colonialism, while also highlighting countervisual practices that foster solidarity and social justice. Mirzoeff advocates for "persistent looking"—a commitment to witnessing and challenging oppressive visual regimes through engaged, critical observation.

Karan J. Odom studies the evolution of complex behaviors, with a focus on sex differences in animal behavior. Her research investigates how environmental and evolutionary pressures shape behavioral traits in both males and females. While traditional studies of sexual dimorphism emphasize male trait elaboration through sexual selection, Odom's work highlights the often-overlooked complexity and elaboration of female behavior. She uses an integrative approach—combining comparative methods, experimental field studies, and neuroendocrine techniques—to explore how environmental changes and social cues influence physiology, gene expression, and behavior. Her work spans both individual lifetimes and broad evolutionary scales across species.

The contribution by Karan J. Odom was collectively written with Michelle J. Moyer, Evangeline M. Rose, Bernard Lohr, and Kevin E. Omland. Odom is mentioned specifically for her elaborate work on the topic.

Ai Ozaki is a Japanese visual artist working with video, photography, ceramics, painting, and text to explore the body's relationship with "the Other"—not only other people, but animals, objects, and even the self. Her practice reflects on the distance and disconnection that often define these relationships, questioning how we try to understand what remains unknowable—like a bird's thoughts or our own internal organs. Through poetic observation and imaginative inquiry, Ozaki investigates the many ways we attempt to bridge these gaps and communicate with beings and systems that are at once intimately near and fundamentally mysterious.

Maria Popova is a Bulgarian writer, curator, and critic based in Brooklyn. She is the creator of *The Marginalian* (formerly *Brain Pickings*), an influential "emporium of ideas" archived by the Library of Congress for its cultural value. Describing herself as a "hunter-gatherer of interestingness," Popova explores art, science, philosophy, and literature to illuminate timeless questions of meaning, beauty, and existence. She is the author of *Figuring* and a champion of children's books for their "absolute sincerity, so deliciously countercultural in our age of cynicism." Her work is a lifelong dialogue between the intellect, the spirit, and the poetic.

Sergio Rojas Chaves is a Costa Rican artist based in San José. With a background in architecture and community development, they work across sculpture, installation, video, photography, and performance. Collaborating with non-human partners such as plants and animals, their practice challenges anthropocentrism through acts of gift, care, and affect. Rojas Chaves explores how biology, ecology, and emotion intertwine, creating works that reframe our relationships with other species. By foregrounding interdependence, they question human exceptionalism and invite viewers to reconsider their place within a shared, entangled world of living beings.

Anna Lowenhaupt Tsing is a Chinese-American anthropologist and writer whose work traces the unexpected life forms and collaborations that emerge in the ruins of capitalism. Her research bridges ecology, global supply chains, and multi-species relations, often focusing on precarious ecologies and resilience in marginal places. In *The Mushroom at the End of the World*, she follows the matsutake mushroom to illuminate how value, survival, and possibility persist amid environmental and economic collapse. Tsing's writing blends ethnography with poetic insight, asking how we can notice and nurture entangled life in the

Anthropocene beyond the myth of progress, and within shared worlds of vulnerability.

Yuri Tuma is a Brazilian multidisciplinary artist based in Madrid whose work explores sonic and queer ecologies through collective practices, sound art, installation, and performance. His practice challenges the human/animal divide imposed by science and Western thought, using sound and collaboration to imagine more interconnected futures. In 2020, he co-founded the Institute for Postnatural Studies (IPS), a center for artistic experimentation from which to explore and problematize postnature as a framework for contemporary creation. At IPS, he practices as its Academic and Artistic Co-Director and co-coordinates its publishing platform, Cthulhu Books, which highlights the political and imaginative power of artistic and academic inquiry to envision alternative relationships between humans, non-humans, and the planet.

Suzanne Walsh is an Irish transdisciplinary artist and writer whose work explores the boundaries between species, systems of knowledge, and forms of communication. Moving between performance, sound, text, and visual media, their practice unsettles dominant narratives by engaging with non-human perspectives, ritual, and language. Through poetic disruptions and speculative modes of address, Walsh reimagines how we relate to the more-than-human world and to each other. Blurring fiction and theory, animality and voice, their work opens space for hybrid forms of being and knowing—inviting audiences into strange, intimate encounters with the unseen, the unheard, and the yet-to-be-articulated.

Inside front cover: John Gerarde, "The Breed of Barnackles," in *The Herball, or Generall Historie of Plantes*, 1597, woodcut, via archive.org.

Inside back cover: Bird nests in the *Journal of the Danish Ornithological Society*, 1913, illustrator unknown, American Museum of Natural History Library.

Collection Metropolitan Museum of Art, U.S.

12, 16, 19, 21, 22, 24, 25, 26, 27, 28, 29, 31, 33, 34, 35, 36, 41, 43, 44, 45, 47, 53, 57, 58, 59, 62, 66, 69, 71, 72, 73, 75, 77, 82, 85, 87, 89

Gallica, National French Library of France

1, 2, 17, 54, 64, 65, 76, 78

Museum of Ethnography, Sweden

32, 63, 80, 95

Nationaal Archief CCO, The Netherlands

4, 11, 60, 83, 97

Wellcome Collection, United Kingdom

3, 15, 55, 93

Rijksmuseum Amsterdam, The Netherlands

5, 6, 13, 39, 48, 50, 61, 74, 88, 94, 98, 100

Smithsonian American Art Museum and its Renwick Gallery, U.S.

7, 9, 10, 37, 38, 51, 56, 67, 68, 79, 90, 91, 96, 101

Smithsonian Libraries and Archives, U.S.

18, 23, 52, 99

Cooper Hewitt, Smithsonian Design Museum, U.S.

20, 30, 70

1 Yves-Marie Le Gouaz, *Anatomie des Oiseaux*, 1778, engraving.

2 Henri Guérard, *Cinq corbeaux*, 1891, lithograph.

3 Peter Ludwig Panum, Bird embryo, 1860, drawing.

4 Jac. de Nijs / Anefo, Common kestrel with three legs at Artis, 1963, photograph.

5 Samuel Jessurun de Mesquita, *Bird of Paradise*, 1914, woodcut.

6 Jan Asselijn, *Threatened Swan*, 1650, oil painting.

7 Édouard Travies, *Hunting: The Common Pheasant*, 19th century, engraving.

8 A prehistoric leg of a juvenile turkey with a complete set of claws, Finnish Heritage Agency.

9 John James Audubon, *Washington Sea Eagle*, ca. 1836–1839, painting.

10 Robert Havell, Jr., Frigate Pelican, from the book *Birds of America*, 1835, engraving.

11 Office for Information and Radio Broadcasting New Guinea, Western crowned pigeon, unkown, photograph.

12 Unknown, from Thailand, Bird, 14th–15th century, earthenware.

13 Ohara Koson, *House sparrows at wisteria*, 1900–1936, woodcut.

14 Corvus octopennatus Daudin, preserved specimen, South Sea, 1800, Naturalis Biodiversity Center / Wikimedia Commons.

15 Unknown, from India, A blue beaked, yellow winged bird, 19th century, gouache.

16 In the Style of Ogawa Haritsu (Ritsuō), Bird, 18th century, watercolor.

17 Henri Guérard, *The Four Dead Crows*, 1888, lithograph.

18 Frederic Augustus Lucas, Skeletal Remains of Great Auk, 1888, photograph.

19 George S. Harris & Sons, Quetzal, from the Birds of the Tropics series (N5) for Allen & Ginter Cigarettes Brands, 1889, lithograph.

20 Unknown, Black Bird with Roses, 1800–1830, drawing.

21 Johann Joachim Kändler, Bird of paradise (quetzal) (one of a pair), 1733, porcelain.

22 Moche artist(s), Peru, Stirrup-spout bottle with owl, 200–500 CE, ceramic.

23 United States National Museum Photographic Laboratory, Bird Hall, Natural History Building, 1956, photograph.

24 Goya (Francisco de Goya y Lucientes), From "The Disasters of War", *The carnivorous vulture*, 1814–15, etching.

25 Unknown, from China, Phoenix, 19th or 20th century, woodblock print.

26 Unknown, from Egypt, Cowroid seal amulet inscribed with a hieroglyphic motif, ca. 1479–1458 BC, steatite.

27 Unknown, from Egypt, Scarab inscribed with the throne name of Amenhotep I, ca. 1525–1504 BC, steatite.

28 Zenu or Sinu artist(s), Peru, Bird Finial, 5th–10th century, gold.

29 Edgar Degas, *Young Woman with Ibis*, 1857–58; reworked 1860–62, oil painting.

30 Unknown, from Japan, Bedcover, 19th century, printed textile.

31 Unknown, from Egypt, Ba-bird, 332–30 BC or later, wood carving.
32 Unknown, from China (Yotkan), Bird, ceramic.
33 Unknown, from England, Angel Scourging a Harpy, ca. 1890, gouache.
34 Suzuki Harunobu, *Young Woman Riding a Phoenix*, 1766, woodblock print.
35 Maya artist(s), from Mesoamerica, Bird ornament, 600–800 CE, shell carving.
36 Native American (Tsimshian), Raven rattle, 19th century, wood carving.
37 Unknown, Elijah Fed by Ravens, early 20th century, wood carving.
38 W. H. Martin, *A good day for ducks in Iowa*, 1909, photocollage.
39 Unknown, from Deccan, Mythological bird, 1750, drawing.
40 Luca Ciamberlano, *The coronation of the imperial eagle under a celestial orb*, 1637, engraving, Municipal Library of Lyon.
41 Unknown, from China, Rank badge with Mandarin duck, 18th century, embroidery.
42 Vicente de Burgos, *Les oiseaux*, 1494, woodcut, Municipal Library of Toulouse.
43 Padihorsiese, Group statue of Thoth-ibis and devotee, 700–500 BC, copper sculpture.
44 Unknown, from Egypt, Relief with a bird, ca. 1353–1336 BC, limestone relief.
45 Unknown, from Egypt, Vulture amulet, 664–332 BC, mineral carving.
46 Aino Alamaa, *Lind*, 1965, sculpture, Tartu Art Museum.
47 Unknown, from Egypt, Black ibis hieroglyph, 100 BC–100 AD, glass plaque.
48 Bernard Picart after Antoine Coypel, *Leda and the Swan*, 1696, engraving.

49 Liebig Cards, Birds of the Congo series, 1955, lithograph, Roubaix Media Library.
50 Jan Brandes, South African Wetland bird, 1786–1787, drawing.
51 Charles Bird King, *Miss Satterlee*, ca. 1830–1839, oil painting.
52 Unknown, Division of Birds Exhibits, Natural History Building, 1910, cyanotype.
53 Unknown, from Egypt, Vessel in shape of bird, ca. 4500–4000 BC, pottery.
54 After H. Bonnart, *Money from my little birds*, 1695–1696, etching.
55 Unknown, from India, Hoopoe perched on a flower stem, 19th century, gouache.
56 Unknown, from North America, Owl Decoy, early 20th century, wood sculpture.
57 Shri Gobinda Chandra Roy, Jatayu trying to prevent abduction of Sita, 1800, woodcut, Cleveland Museum of Art.
58 R. W. Martin and Brothers, Tall bird, 1896, ceramic.
59 Unknown, from Egypt, Seal in shape of a duck, ca. 1550–1450 BC, steatite.
60 Rob Bogaerts / Anefo, Television "bird" interviews the politician Onno Ruding, 1988, photograph.
61 Alexander Cranendoncq, *Ziet hier al weer een nieuwe prent; Die m'om uw leerzaamheid u schenkt*, 1819–1840, woodcut.
62 Unknown, from the Netherlands, Bird, 15th century, metal sculpture.
63 Unknown, from China (Yotkan), Bird, unknown, ceramic.
64 De la fabrique de Pellerin, Various birds No. 2, 1838, colored woodcut.
65 Photographic agency Rol, Marché aux oiseaux, 1928, photograph.
66 Unknown, from Egypt, Tell el-Yahudiya, ca. 1700 BC, ceramic.
67 Unknown, Woman and Tree with Birds Nesting, 20th century, sculpture.
68 Blanche McManus Mansfield, *The True Mother Goose*, 1895, lithograph.
69 Unknown, from Japan, Duck figure, 19th century, ceramic.
70 Harvey Smith, Design for Wallpaper and Textile: Birds, 19th century, drawing.
71 Sarah Furman Warner Williams, Phebe Warner Coverlet, 1803, textile.
72 Unknown, from Iran, Figure of Harpy, 12th–early 13th century, ceramic.
73 George S. Harris & Sons, Curlew, from the Game Birds series for Allen & Ginter Cigarettes Brands, 1889, lithograph.
74 Kamisaka Sekka, *White Phoenix*, 1909, woodcut.
75 Unkown, from Egypt, bird-shaped tool, ca. 3300–2900 BC, flint.

76 Claude Gillot, *The Birds*, 1719, engraving.
77 Francesco Bartolozzi, Prometheus, naked, chained to a rock, looking up at an eagle standing over him, 1795, painting.
78 Gabriel Salmon, *A Rooster and a Hen in the Henhouse*, 1520, woodcut.
79 Helen Hyde, *My Neighbors*, 1913, woodcut.
80 Unknown, from North Korea, Bird trap, 1937, wicker.
81 Lucas Cranach the Younger, Bird study, 1560, drawing, Albertina Museum, Vienna.
82 Unknown, from Greece or southern Italy, Terracotta oil flask with a bird, 4th century BCE, ceramic.
83 Eddy Posthuma de Boer / RVD, Princess Beatrix with a bird in a cage, 1964, photograph.

IMAGE CREDITS

84 Max Rosenthal after Henry Louis Stephens, Taylor Bird, 1851, lithograph, National Portrait Gallery, Smithsonian Institution.
85 Unknown, from Vietnam, Ewer in the Form of a Phoenix, ca. 15th–16th century, ceramic.
86 Max Rosenthal after Henry Louis Stephens, *Jolly Old Cock*, 1851, lithograph, National Portrait Gallery, Smithsonian Institution.
87 Unknown, from Egypt, Ibis, 664–30 BC, metal sculpture.
88 Melchior d' Hondecoeter, *Hunting booty at a landing*, 1678, oil painting.
89 George S. Harris & Sons, Nightingale, from the Song Birds of the World series for Allen & Ginter Cigarettes, 1890, lithograph.
90 Helen Hyde, *Feeding the Geese*, 1918, etching.
91 Kerr Eby, *Shark Rock*, 1941, etching.
92 Jacques Fornazeris, Peacock, 1594, etching, Municipal Library of Lyon.
93 Unknown, from India, A bird forming a tughra (cipher), woodcut.
94 Mathieu Lauweriks, Swan behind fence, 1935, woodcut.
95 Unknown, from India, Pinjra, Birdcage, 1909, wood work.
96 Unknown, American, Rooster Weathervane, 19th century, metal sculpture.
97 J.D. Noske / Anefo, Yellow-billed stork at Artis, 1959, photograph.
98 MH, Owl dressed as a soldier, 1500–1549, woodcut.
99 United States National Museum Photographic Laboratory, Bird Hall, Natural History Building, 1956, photograph.
100 Jacques de Fornazeris, *Jay*, 1580–1590, etching.
101 Unknown, Calligraphic Rendering of an Eagle, 19th century, ink drawing.
102 Unknown, Western Siberia or Volga/Kama region, Bird, 4th–11th century AD, bronze, Los Angeles County Museum of Art.

PERMISSIONS

with a Bird,
A Reader on Avian Kinship

With contributions by:
John Berger • Ignace Cami • Sara Sejin Chang (Sara van der Heide) • Monika Czyżyk • Bryony Dunne & Suzanne Walsh • Daniel Godínez Nivón • Daisy Hildyard • Manjot Kaur • Natalie Lawrence • Marianne Elisabeth Lien • Michelle J. Moyer & Evangeline M. Rose & Bernard Lohr & Karan J. Odom & Kevin E. Omland • Nicholas Mirzoeff • Ai Ozaki • Maria Popova • Sergio Rojas Chaves • Anna Lowenhaupt Tsing • Yuri Tuma

Editor: .... Marjolein van der Loo
Graphic design: .... Studio Yannick Nuss
Copy edited by: .... Harriet Foyster
Published by: .... Onomatopee
Typefaces: .... Arial, Bookman, Cooper Black, Fourteen64, Frankfurter, Hobo, IM Fell, Mikadan, New Rail Alphabet, Whimbrel Caps by Alex Tomlinson.
Paper: .... Caribic (Chromgelb)
Printed by: .... AS Printon
Edition: 4000

Onomatopee
Lucas Gasselstraat 2A
5613 LB Eindhoven
The Netherlands
www.onomatopee.net

cultuur eindhoven

mondriaan fund

ISBN: 978-94-93382-02-2 Onomatopee #262

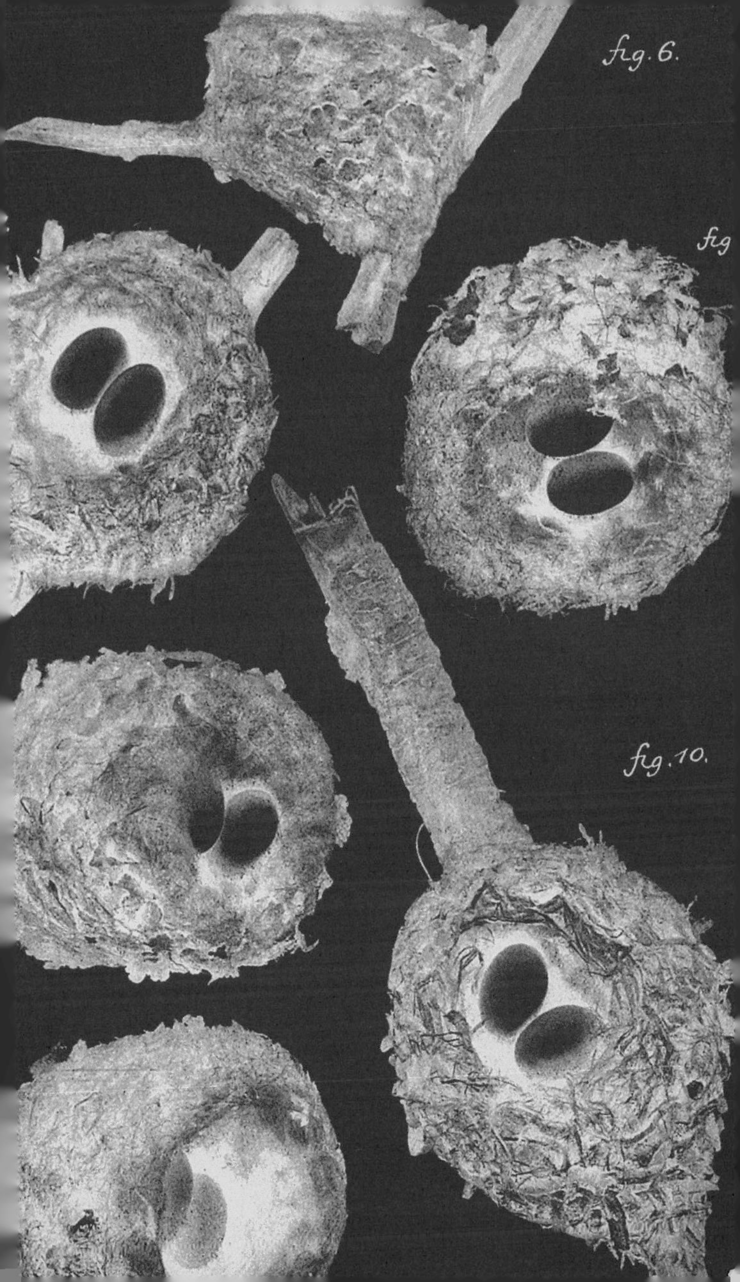
fig. 6.
fig
fig. 10.

With a Bird, A Reader on Avian Kinship invites readers into an expansive, cross-disciplinary conversation about how we live with and think alongside birds. In a time of climate breakdown and ecological grief, this book offers birds not as metaphors or curiosities, but as kin—creatures with their own histories, desires, and forms of knowing. Spanning speculative fiction, ancestral memory, critical ornithology, personal essay, and visual art, its contributions explore the fragile, often overlooked relationships between humans and birds across myth, science, migration, and dream.

Through listening and attention, the book explores how birds shape landscapes, signal planetary change, and offer new ways of understanding time, voice, and relation. Contributors draw on decolonial, feminist, and ecological practices to unsettle dominant narratives and invite forms of care, reciprocity, and repair. From the mimicry of the lyrebird to the silence of vanished species, from winter dreaming to co-domestication, each chapter gestures toward multispecies futures grounded in presence and poetic attention.

This reader, both a continuation of an exhibition and a gathering of distinct voices, becomes a spell—woven from memory, sound, and image—that reimagines kinship in flight.

ISBN: 978-94-93382-02-2